Contents

Introduction 5

Chapter 1 Electricity 10

Chapter 2 Magnetism 41

Chapter 3 Energy 56

Chapter 4 Forces 64

Chapter 5 Light 102

Chapter 6 Sound 131

Chapter 7 Earth in space 145

Glossary 167

Index 172

Icon key

Information within this book is highlighted in the margins by a series of different icons. They are:

Subject facts
Key subject knowledge is clearly presented and explained in this section.

Why you need to know these facts
Provides justification for understanding the facts that have been explained in the previous section.

Common misconceptions
Identifies and corrects some of the common misconceptions and beliefs that may be held about the subject area.

Vocabulary
A list of key words, terms and language relevant to the preceding section. Vocabulary entries appear in the glossary.

Amazing facts
Interesting snippets of background knowledge to share.

Teaching ideas
Outlines practical teaching suggestions using the knowledge explained in the preceding section.

Questions
Identifies common questions and provides advice on how to answer them.

SCHOLASTIC

The Primary Teacher's Guide to **Physical Processes**

• Key subject knowledge • Background information • Teaching tips •

SCHOLASTIC

MIX
Paper from responsible sources
FSC® C007785

Book End, Range Road, Witney, Oxfordshire, OX29 0YD
www.scholastic.co.uk

1 2 3 4 5 6 7 8 9 2 3 4 5 6 7 8 9 0 1

British Library Cataloguing-in-Publication Data
A catalogue record for this book is available from the British Library.

ISBN 978 -1407-12786-6
Printed and bound by CPI Group (UK) Ltd, Croydon, CR0 4YY

Author
Neil Burton

Commissioning Editor
Paul Naish

Development Editor
Emily Jefferson

Proofreader
Kate Pedlar

Indexer
Penny Brown

Illustration
Garry Davies and Stephen Lillie

Icons
Tomek.gr

Series Designers
Shelley Best and Sarah Garbett

Acknowledgements

I dedicated the first edition of this series to 'my boys' - so this one's for 'my girls', Martine, Chantelle, Kerry, Helen and Alyssia.

Many thanks to Jill Trevethan, my colleague, and James, my son, for reading through drafts and commenting - I still wish to claim that all errors are entirely my own work!

The publishers gratefully acknowledge permission to reproduce the following copyright material:

Pan Macmillan Ltd for use of an extract from The Hitchhiker's Guide to the Galaxy by Douglas Adams, text © 1979, Serious Productions Ltd (1979, Pan Books).

Every effort has been made to trace copyright holders for the works reproduced in this book, and the publishers apologise for any inadvertent omissions.

Physical processes

What is a primary scientist?

There is an important distinction to be made between scientists and primary science teachers. A scientist follows a particular strand of science as far as possible in order to explain it. A primary science teacher tries to understand the scientific links between things around him or her, and to help others achieve a better understanding of them. A good primary science teacher is interested in helping children to understand ideas, explain them and link them together. The words 'explore', 'investigate' and the phrase 'Why do you think that happened?' are much more likely to come from such a teacher than 'Learn this by heart' or 'You don't need to understand it, just remember it!'

How can this book help?

Having taught primary science within primary schools, initial teacher education and continuing professional development over the last thirty years, I am convinced that there has been a steady improvement in the level of teachers' background science *knowledge*. I am less convinced that this has been entirely transferred to an *understanding* of the concepts at a level at which teachers are secure and confident. To have a level of understanding that allows you to pass on 'facts' with a fair degree of accuracy is not enough. You need to have a 'feel' for the subject that will allow you to appreciate where the children are 'coming from' and enable you to help them learn more effectively by identifying the 'next steps' they need to take or the deficiencies in past learning that their misconceptions reveal, signalling a need to 'revisit' before moving securely on.

This book attempts to avoid a 'GCSE science – revisited' approach. GCSE textbooks serve a particular purpose well: to get pupils through a particular type of science exam. Primary school teachers have a very different need: to develop a clear and focused understanding of the science in order to be able to teach it effectively to others.

The idea of 'keeping one page ahead in the book' is now, surely, long gone. Teaching is much more than the passing on of facts for future regurgitation by a new generation of learners. Before teaching comes assessment and planning. Initial assessment for learning is needed to reveal the children's ideas, to find out what they know and what they don't, what they think they know and what they don't realise that they do! Once you move on to finding out what their level of understanding is, things start to become less clear, but this is necessary in order to find out where the learning should start. After assessment comes planning: deciding on the ideas that you want the children to develop and the challenges, tasks and questions that you expect will get them there. Finally, there is the delivery: the teaching and learning that will see the plans come to fruition. Such a process can be embedded in the science curriculum for a key stage, or can be part of a pupil/teacher interaction lasting a few seconds. The key is understanding where the children are, where they need to go next and how to get them there – that is, understanding the science.

This book does not try to give you all the scientific facts you will ever need as a teacher. That's impossible: science is changing far too quickly. What it will attempt to do, largely through clear explanations, models and analogies, is to help you to reach a better understanding of the science – to help you visualise what is happening. Some of the science has been simplified in an attempt to clarify the ideas, but every effort has been made to ensure that the material is scientifically correct.

Structure of this book

Each area of science is broken down into a few **key concepts**. These are the ideas used as a focus for the development of explanations and examples. Together they form the basis of the elements within each strand of science of which children (and consequently you) need to develop a secure understanding.

A **concept chain** is included to provide a clear indication of progression and development. The chain goes beyond Key Stage 2, in order to help you see where the children are going. Scientific understanding is a never-ending quest; but an appreciation of where a child is on the learning continuum allows previous points to be reinforced and the next ones to be explored. With any scheme of work, a sense of place, pace and direction is essential for all concerned.

The **subject facts** provided have three main purposes:

- To help you understand the ideas so that you can teach them effectively. (The explanations will often go beyond what you need to teach at Key Stage 2.)
- To show where the children will be going next with their learning.
- To make it easier for you to identify where the children have developed misconceptions.

Connected to this is a list of the important technical **vocabulary** that will be required to teach and understand these key ideas. It is especially important for the children to realise that some words used in general conversation have particular and explicit meanings in science, and to begin to appreciate when they need to apply this scientific usage.

Children are generally fascinated by **amazing facts**. Each scientific theme includes a few with which you can impress the class. They are mostly chosen to give some idea of the scale or extremes within the topic.

An understanding of some **common misconceptions** and what you might do about them is important in both assessment and planning. If you are aware of ways in which children might hold misconceptions, you can plan activities in such a way that children holding these ideas can be identified. Focused teaching can then be effectively employed. This section offers advice on identifying and rectifying pupil misconceptions.

Some common **questions** that children frequently ask about the phenomena and processes within particular areas of science are presented, together with suggestions on how they can be answered to give children an understanding of the science involved. However, the answers may not always provide a full explanation!

Further suggestions for practical **teaching ideas** are provided, with a focus on particular approaches (exploring, investigating, sorting and classifying) and on the use of ICT to enhance learning. Where possible, specialist science equipment is avoided. This helps to link the science to familiar and practical contexts. However, in some cases particular equipment is necessary for measuring or collecting, or is required for health and safety reasons. The ICT activities depend on a certain level of hardware and software availability: word and data processing facilities; the ability to capture sound and images and incorporate them into presentation software; remote data capture (for example, temperature probes). The use of the internet should be encouraged, but because of the transitory nature of many websites, research prior to any classroom use is advised.

Approach to teaching and learning

This book leans towards a particular learning theory – without, I hope, slavishly following it. The constructivist approach to the teaching of science has developed over many years and is now acknowledged as a highly effective way of ensuring that children develop scientific abilities on the basis of a secure foundation in scientific concepts. Constructivism attempts to avoid teaching children an understanding of a scientific process or phenomenon while, at the same time, allowing them to continue to hold a conflicting conception. It is easy for a child to believe two very different explanations for the same concept concurrently. They may have been told one explanation at school and use it to respond when the teacher asks, but hold another, arising from their empirical observations, which they use the rest of the time (for example, 'plants go to sleep at night' based on the observation of flowers closing).

To adopt a constructivist approach, the teacher must first determine the area of science within which the children will need to work (from the curriculum or the school's scheme of work). This will help the teacher to identify the range of ideas and misconceptions that the children might hold. Next, the teacher needs to identify a stimulus or context that will help the children to focus on the topic by engaging their interest in something familiar. The next three stages depend on the children. In the first stage, the stimulus or context is used to elicit the children's ideas, with the teacher questioning and probing to discover the extent of their understanding. Based on this assessment, the teacher plans a range of highly focused activities, building on the children's questions and suggestions to reinforce correct ideas and challenge misconceptions. In a final assessment and reflection stage of the cycle, the children are asked to consider how their understanding has changed. The teacher acknowledges progression and through this assessment for learning identifies the next steps to be taken in the next cycle of learning. There are many different 'levels' of cycle in operation here – perhaps annual, as the topics to be taught in the year are reconsidered, down to the momentary, as the pupil and teacher interact and the teacher offers different responses in reaction to what the learner does or says.

This approach is about constructing a secure understanding of science based on broad and firm foundations, rather than requiring the children to retain transitory knowledge, which they are unable to link to any pre-held ideas. However, there is a downside. This approach tends to take longer, both in the meeting of individual learning objectives and in terms of the coverage of content, than a more piecemeal approach. Although 'finding out where the children are starting from' is an essential prerequisite, it could mean that the teacher is attempting to help the children progress from many different starting points. Applying probing questions to a whole class cannot be done both accurately and quickly – though if you have found a way, please get in touch! The recording of assessments is also very time-consuming, although it can be very satisfying to look back and register the progression. The teacher will need to group children with similar ideas and find an efficient way of charting misconceptions and progress.

Not only is constructivism a highly effective way of teaching children, it also encourages a reflective and professional approach from the teacher. It leads the teacher to examine and analyse every step of the teaching; this will aid the development of a better understanding of the individual children and their learning, as well as of the scientific ideas and the means of communicating them.

Chapter 1

Electricity

Key concepts

Concepts concerning electricity are quite wide-ranging at the primary level, covering the nature of electricity, circuit construction, applications of electricity and safety issues. The key points are:

1. Electricity has particular, well-defined properties.
2. It is the power source for many household appliances.
3. It is potentially dangerous, but its risks can be minimised.
4. It is able to flow around an unbroken circuit.
5. Switches can be used in a circuit to make different components do different things.
6. Electricity will affect components arranged in series and in parallel circuits differently.

Electricity concept chain

It is important to be aware of how these concepts can be developed in teaching. The following is *one* way in which the progression can be described. It goes up to Key Stage 3, because it is necessary to know where the children will be going next. To demonstrate your own understanding of the concepts, it is useful to produce your own concept chain.

KS1

Many everyday items, including toys and appliances, use electricity. Some items use mains electricity and some use batteries; some are able to use both. Mains electricity is potentially dangerous if the correct safety procedures are not followed. Simple circuits can be constructed from batteries and bulbs, connected together using wires. Batteries and bulbs work best if their voltages are closely matched. Buzzers will only operate if connected a particular way around. Most metals are good conductors of electricity. Electrical devices in a circuit will only work where there is a complete conductive path.

KS2

Devices in a circuit can be controlled by switches which make and break the circuit. A completed circuit can be explained in terms of the continuous flow of electricity through it. Increasing or reducing the voltage of the battery used in a circuit will affect the operation of devices in that circuit (bulbs will become brighter or dimmer). Placing an additional device in the circuit will reduce the output of each device (for example, if a second bulb is added, both bulbs will be dimmer than the first was alone). A *series* circuit is one where there is only one path for the flow of electricity to take to complete a circuit. A *parallel* circuit is one where there is more than one path for the flow of electricity to take to complete a circuit. Devices connected in parallel act as if they were in circuits on their own. Batteries connected to a parallel circuit with two devices will expend their energy more quickly than if the same devices were connected in series. A circuit diagram, with symbols for devices, can be used to represent a circuit. It can also be used as a basis for the construction of a circuit. Placing a variable resistance in series with a device will allow the output of that device to be varied (for example, making a bulb less bright).

KS3

The flow of current throughout a series circuit is the same. The sum of the voltages measured at each device in a series circuit will be the same as the output of the battery or power supply connected to the circuit. *Voltage* (V) is a measure of potential difference (electrical pressure) between the two terminals (or electrodes) of a battery, measured in volts (V). Electrical *current* (I) is a measure of the flow of electricity through a circuit, measured in amps (A). The resistance to the flow of electricity in a conductor depends on the thickness, length, material and temperature of the conductor. Electrical *resistance* (R) is measured in ohms (Ω), and has a heating effect on the surrounding material. Electrical current, resistance and voltage are linked by the relationship $V = IR$. *Power* (P), the rate at which electrical energy is used, is related to current and voltage by the relationship P = IV. A conductor is a material in which the electrons can move freely from atom to atom. Electricity is a flow of electrons in a particular direction. The flow of electricity causes an electromagnetic field, which can be used to repel and attract a permanent magnet and so cause movement. Movement of a coil of wire in a permanent magnetic field will cause a flow of electrons in the coil.

Concept 1: What is electricity?

Subject facts

Nature of charged particles

In order to begin to understand what electricity is, it is necessary to know some things about the make-up of physical substances, or matter. Matter is made up of small particles, or atoms. A useful model of an atom is a central nucleus containing positively charged particles, surrounded by a 'cloud' of negatively charged particles, or electrons. Normally, in any one atom, the positive and negative charges will balance; but sometimes they can become unbalanced, leading to a build-up and flow of electrons that we call electricity.

'Static' charge

Some insulators (materials that do not readily allow electricity to flow through them) can build up what is termed a 'static' charge through friction. The effect of rubbing an inflated balloon against hair is a good example. Extra electrons are collected by the balloon, giving it an overall negative charge. The attraction of opposite charges allows it to 'stick' to a non-conducting, positively or neutrally charged surface such as a wall or ceiling, or to attract fine hair and make it stand on end.

Sometimes, when the charge difference built up is very large, it may balance out by causing all of the electrons to be discharged very quickly. As you get out of a car or walk across a carpet, you may receive a 'shock' – a rapid discharge of electrons. Because of the movement of water droplets within thunder clouds, huge charges can build up inside them. When these discharge against oppositely charged parts of the cloud or against the ground, the energy of the charge literally blasts the air apart in creating a path for the discharge. When this happens, the explosion can be seen as a flash of light and heard as thunder. A spark from a gas cooker ignition system has the same effect, though on a much smaller scale.

Electrical flow

The electrical flow associated with **battery** and mains electrical equipment is very different from a static charge. Electricity of this kind flows through **conductors** (materials that allow electricity to flow through them), which are usually metals. Metals have a particular structure that means the electrons are not securely attached to any one atom, allowing them to 'wander' from atom to atom in a random way; they can thus become part of an electrical current. This form of electrical flow is very different from the electromagnetic radiation of light or radio waves (see Chapter 5, page 102).

Resistors

In a relatively thick piece of a good conducting material, such as a cable, the resistance to the flow of electrons is minimal over the distance the electricity is flowing. The thinner or poorer the conductor, the more the flow is restricted. As the electrons attempt to flow through the material, the resistance to their flow causes a kind of friction that makes the material heat up. Sometimes a **resistor** will heat up so much that it begins to glow.

Resistance can be useful! Electrical resistance is used to generate heat and light in filament light bulbs, electric kettles, electric fires and hairdryers. The special part of these devices which resists the flow of electricity and heats up is called the 'element'.

Electricity generation

Electricity is generated from a number of primary energy sources, all of which produce movement that operates a dynamo. In a dynamo, a coil of wire is turned within a strong magnetic field; this causes an alternating current to be induced in the coil, and so electricity is produced. Most generation systems burn fuel (oil, coal and gas) or use a nuclear reaction to heat water;

the steam is used to turn turbines that drive the dynamos. Hydroelectric power, windmills, tidal and wave generators use movement to turn the turbines directly, and are therefore more environmentally friendly.

Electricity transmission

Electricity is distributed around the country, from power stations to consumers, by the National Grid. This is a network of 'power lines' which operate at charge differences (voltages) of up to 40,000 volts. As domestic systems in this country only use 220–240 volts, the voltage has to be 'stepped down' in transformers for local supplies. By transporting electricity around the country at very high voltages, there is less energy loss due to electrical resistance.

Batteries

Batteries contain chemicals which 'store' electrical charge. When connected into a complete circuit with a conducting material, the battery causes all of the wandering electrons in that material to move in a particular direction. This effect is known as direct current (DC). One way of imagining a battery is as a box with a compressed spring in it (when it is fully 'charged'). When the battery is connected into a complete circuit, the spring slowly releases its energy to push the electrons around the circuit. When it is completely decompressed, it will have used all of its stored energy and the battery will be 'flat' or 'dead'. With rechargable batteries, visualise the spring being recompressed, so that energy can be released once more. You could also visualise recharging a battery as compressing air by blowing into a balloon – once the balloon is fully inflated (or 'charged') it is a store of (potential) energy.

Mains electricity

Because of the way it is generated (see page 13), mains electricity supplies an alternating current (AC). The direction of the current changes back and forth very rapidly (in the UK, 50 times a second). This means that the electrons are made to go first in one direction and then in the opposite direction. This is like the difference between a handsaw (which goes backwards and forwards) and a chainsaw (which always goes in the same direction). DC is neither more nor less powerful than AC.

Why you need to know these facts

Strictly speaking, children at KS1 and KS2 need only to consider the effects of electricity without being too concerned about its nature. Given the difficult concepts involved in theories about the nature of matter, this approach is sensible and makes the subject more accessible to children. However, it is important that teachers have some appreciation of the nature of electricity, so that they are able to understand the progress the children will be expected to make later on. If the children gain a secure foundation through an understanding of the applications of electricity in their primary science work, they will readily make the later transition to higher levels of understanding.

Vocabulary

Battery – a case containing chemicals that react to cause a flow of electrons when connected into a complete circuit.
Conductor – a material that allows the free flow of electricity through it (that is, the free movement of electrons between atoms).
Resistor – a device that restricts the flow of electricity in a completed circuit. Some resistors, such as a bulb or a motor, use the 'friction' of the restricted flow to produce light, heat or movement.

Amazing facts

- The electric eel of South America can generate an electric charge of up to 650 volts – enough to stun a horse as it wades through a river.
- Lightning flashes can be up to 32km long and 3 metres in diameter. In Britain, there is an average of around 3.5 strikes per square km per year.

Common misconceptions

Note that much of the teaching of electricity at KS1 and KS2 has to do with the practical issues of making circuits, using electrical devices and observing the effects. There is no need to focus on the children's understanding of the nature of electricity; but an overly naïve interpretation of electrical effects can lead to misconceptions which will cause problems later on.

Electricity comes from shops.

Well, batteries do – and you can buy meter cards in shops. However, children should not think that mains electricity is carried home in a bag with the groceries. Ask questions such as: *How could it get from the shop to the socket? What do we need to do to a hairdryer to make it work?* Establish that mains electricity comes to our homes in cables, and explain about sockets and plugs.

Questions

What is electricity?

This one is the biggy! You need to judge what the child is capable of handling, and what you are able to explain. At this level, the emphasis should be on the effects of electricity, so try to explain it in term of what it does: *It's what makes a bulb light up*. If you are really pushed by a more able Year 6 child, it is better to admit uncertainty and direct him or her towards a textbook than to attempt to give an insecure explanation. To understand the 'flow of electrons' idea, children will need to some understanding of the nature and structure of matter and how this relates to electrical conduction. This will, no doubt, result in more questions than answers – which is the way that all good science develops. Take comfort in the fact that very few people could say with confidence that they really understand electricity. The key thing for you to ensure is that the children get to the next step without carrying too many misconceptions with them.

Teaching ideas

Conductivity (testing and sorting)

With KS1 (or younger) children, construct a simple circuit with a battery, a bulb and a break in it. The children can then sort a given range of materials into ones that allow the bulb to light when they are used to fill the break in the circuit, and ones that don't. This can be extended by asking the children to test other things from around the classroom. ***Make it very clear that they must not touch anything that has to do with mains electricity (either the sockets or the equipment).***

Concept 2: How do we use electricity?

Subject facts

This should be considered on two levels. Firstly, what things use electricity as their main power source? Secondly, what form of electricity do they use?

Electrical equipment

Most households have a vast range of electrical equipment that they have come to rely on, from a tumble-dryer and a TV to alarm clocks and light bulbs. The identification of things that use electricity can be based on looking for cables coming out of the back that attach to wall sockets. But you can't always see the cables; and even when you can, it doesn't mean that electricity is the main power source – electrical switching often controls gas cookers and central heating systems.

Many items of equipment have the option of using either mains or battery. For example, a radio may be designed to run on batteries and also contain a device for transforming mains AC to DC at the correct voltage. Battery-powered toys use electricity to produce movement, light or sound.

Electrical power

Power is measured in watts (W), after James Watt (1736–1819) of steam engine fame. It is a measure of how much energy is being used in a given time: one watt equals one joule (J) per second, so a 60W electric light bulb uses 60J per second (see page 57). The greater the power rating (wattage) of an electrical item, the more energy it uses in a given time. An 18W 'energy saver' light bulb gives out as much light as an old fashioned 100W tungsten filament bulb but only requires 18% of the energy. These old filament bulbs only managed to convert about 2% of the electrical energy that they received into visible light – the new 'energy saving' fluorescent bulbs convert about 22% into light – still not very efficient! Generally, a 2kW kettle will boil a given amount of water twice as fast as a 1kW kettle.

Fuses and circuit breakers/trip switches

Fuses protect electrical equipment from overheating by only allowing a certain amount of electricity to flow through them at any time. They break the circuit if the **current** (the flow of electricity) becomes too great or there is a surge in the **voltage** (electrical pressure). Fuses are rated in amps. For domestic 13-amp plugs, they usually come in 3A, 5A and 13A ratings. The electrical power rating of a piece of equipment tells you what fuse you need to protect it with. The voltage of mains electricity in this country is between 220V and 240V. Power (in watts) is the voltage multiplied by the current (in amps). A 2kW kettle requires a current of about 8A (2000 ÷ 240), so it needs a 13A fuse. All new electrical equipment should be sold with a plug attached that has the correct fuse inserted. Most homes now have circuit breakers rather than fuse boxes – if a piece of electrical equipment has a fault (a bulb 'blowing' will sometimes cause this), the circuit is switched off and will need to be reset (the switch turned back on) at the electricity distribution box.

Why you need to know these facts

Children will be aware of electricity as the most commonly used energy source; but they need some understanding of its uses to be able to use it safely. (Also see Concept 3, page 20.)

Vocabulary

Current – the rate of flow of electricity, measured in amps (A).
Power – how much energy is used in a given time, measured in joules per second or watts (W).
Voltage – the difference in potential between two parts of a circuit, 'pushing' the current. Voltage is measured in volts (V).

Common misconceptions

Commercial imitations of non-electrical objects, such as candle-shaped light bulbs or 'coal-effect' fires, may confuse children who have never seen the real thing and hence assume that the latter requires electricity!

There can be some confusion about rechargeable electrical items such as hand mixers or remote-controlled toys. Children may think that the electricity is coming directly from the mains. In reality, mains electricity is used to recharge the special batteries stored in the device.

Teaching ideas

Mains and battery (sorting)

The children can use an old catalogue of electrical goods to make a collage of two sets: mains-powered and battery-powered equipment. They could show an intersection (objects that use both) if appropriate.

Alternatives (matching)

The children can return to the catalogue, cutting out and mounting five or so examples of equipment which make different uses of electricity (such as a battery shaver, digital alarm clock, hairdryer, food mixer or a fan heater), then find non-electrical alternatives (a wet razor, clockwork alarm clock, towel, hand whisk or a gas fire).

Concept 3: What can electricity do to you?

Subject facts

Electrical safety

The human body, though its muscles and nerves are controlled by electrical pulses, is not a good conductor of electricity. Neither is the air. Lightning is a good example of what electrical charges can do if enough energy builds up: the discharge blasts its way through the air.

Normal, non-rechargeable batteries up to 4.5V can be used safely in the classroom, as long as the children do not attempt to eat them or throw them at each other (the chemicals are toxic). I have known children to complain of getting 'shocks' when working with circuits; but in every case, this was due to them catching their fingers with crocodile clips or ends of wire.

Rechargeable batteries

These can be very cost-effective when used in working models or toys where the circuit has been tested and is correct. When experimenting, children may accidentally 'short circuit' the battery by connecting a wire directly from one terminal to the other. This will allow the charge to flow very easily, leading to a build-up of heat; rechargeable batteries can discharge very quickly, given the opportunity. The heat will either manifest itself in the wire (it may melt the plastic covering) or in the battery (the chemicals may expand and squirt out). Whilst most rechargeable batteries have means to prevent this happening – you can't be too careful!

Mains electricity

As a general rule, children should *not* operate mains equipment, especially at the wall socket. Having said that, items of equipment such as computers and glue guns are quite frequently placed in the hands of children, following appropriate safety training. All electrical equipment must be checked for safety on a regular basis by a qualified technician, and classroom ring mains (the wall sockets) should be protected with earth-leakage breakers which will turn the electricity off as soon as it is detected leaking out of the system. There will be a reset button somewhere nearby, usually quite high up on a wall. Check for any problems before you attempt to reset; if in doubt, call for an electrician.

Why you need to know these facts

Children are very familiar with equipment that uses electricity, and are often allowed to operate it. In most cases this is quite safe, but familiarity can breed contempt for safety issues. They need to be made aware of the potential dangers of working with electricity, and how these hazards can be minimised.

Teaching ideas

Safety session

Use resources available from your local electricity company to highlight safety issues. The children can watch safety videos, circle hazards on posters and devise safety slogans.

Concept 4: Electrical circuits

Subject facts

Matching components

Before constructing an electrical circuit, you need the right components. It is important to use batteries and bulbs that, as far as possible, have similar voltage ratings. The 1.5V battery (either an AA or a C type) and 1.5V bulb are a good starting point. The voltages of batteries are quite easy to read, but those of bulbs are more difficult. On the metal casing just below the glass bubble, you should be able to read the voltage (this may require a magnifying glass). The metal terminals on the battery allow you to light up the bulb without using any connecting wires.

If you try using a 1.5V battery with a 3.5V bulb, you will find that the bulb lights very dimly. Using two 1.5V batteries in a 'battery box' with a 1.5V bulb will make the bulb very bright for a very short period of time, until the filament gets too hot and burns out. You can think of voltage as 'electrical push': the higher the voltage, the greater the push. A 1.5V bulb needs a 1.5V 'push' from a battery to make it work properly. If you push it too hard, it will break. If you don't push it hard enough, it won't have enough energy to light up well.

Batteries and bulbs

If you start with just a battery (1.5V) and a bulb (1.5V) and a single piece of connecting wire, you can see all of the important connections that need to be made for a larger, more complex circuit to work. Identify the two terminals at either end of the battery. On a screw-fitting bulb, the terminals are the pimple on the bottom and the outer screw casing itself. To make the bulb light, each of the battery terminals must make contact with a bulb terminal (see Figure 1).

This is not the most convenient way to hold a circuit together, so the connection can be maintained with wires. Cut each wire to length, remove the outer covering at the ends using wire strippers, and twist the strands together for safe and efficient use. Now use the wires to connect the terminals of the battery and the bulb (see Figure 2a). This is still fiddly; so place the bulb in a bulb holder, connect the wires to the terminals, and use adhesive tape to fix the wires onto the battery (see Figure 2b). This is an effective circuit.

Figure 1
A simple circuit

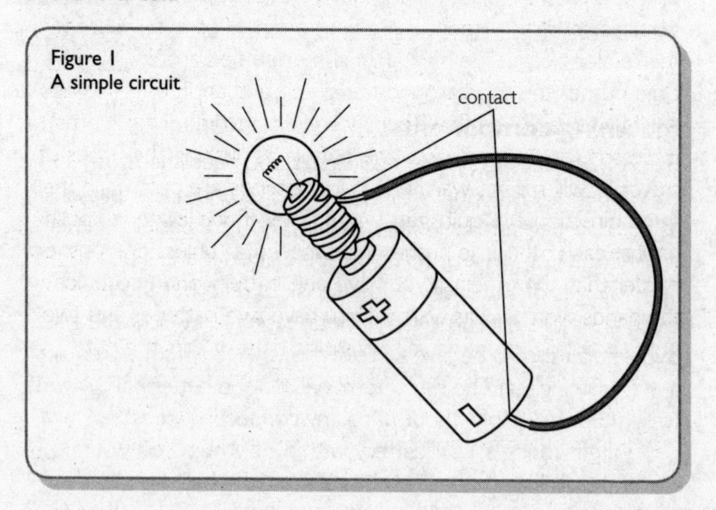

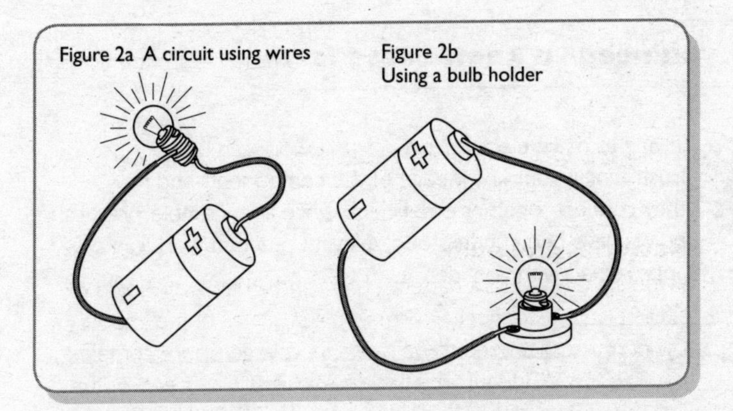

Figure 2a A circuit using wires

Figure 2b Using a bulb holder

Circuit models

Explanatory models for the flow of electricity need to be clear, well-focused and within the experience of the pupils.

Two models are commonly used for teaching about electricity:

- In the 'Smartie®' model, children (acting as 'electrons') are given Smarties® by the teacher. They eat them, climb over a chair and return to the teacher. This expresses the idea that batteries provide energy (the Smarties®) for the electrons to do work within a **circuit**. But it leaves unanswered questions such as 'How does the electron know that there is only one chair in the circuit and it won't have to save half a Smartie® for later on?'
- The 'skipping rope' model, though more simplistic, is more versatile: four sticks (pencils are fine) are held up in a square (a metre along each side is ideal), with a rope tied around them. A child pushes the rope around the circuit, and another child grips it lightly to resist the flow (being careful to avoid friction burns). It doesn't matter which way the 'battery' is pushing: the hands of the 'bulb' will still get warmer. If two 'batteries' are pushing in the same direction, the 'bulb' gets warm faster. If two 'batteries' push against each other, the pushes will cancel out (unless one pushes harder than the other). If you have one 'battery' and two 'bulbs', the hands won't get as warm. If you have two batteries and two bulbs, it will be the same as having one battery and one bulb.

This model is useful for helping children to develop an understanding and visual image of what happens in a circuit. It holds up fairly well in practice; but as with all models, you should recognise its limitations and not take it too literally.

Why you need to know these facts

Trying to make electrical circuits that work can be very frustrating if your knowledge of the components and how they connect together is deficient. Once a few simple, practical points have been understood, relevant practical work can be approached with confidence.

Vocabulary

Circuit – a complete path between the two terminals of a battery, made with materials that conduct electricity.

Common misconceptions

Different-coloured wires have different properties.

Wires are covered in different-coloured plastic sheaths to make each one easier to identify in a tangle of wires. However, this is not always practical: in a circuit where bulbs are connected in parallel, trying to use a different-coloured wire for each bulb might require too many colours. In practical terms, it is the metal inside the plastic that is the important factor when making connections. This is particularly important when distinguishing between the live (brown), neutral (blue) and earth (green and yellow) wires used in household appliances.

Some devices (such as buzzers) contain *diodes* that only work when connected to the battery one way round. In these cases, it is conventional to attach a red wire to the positive terminal.

A bigger battery has more electricity in it and will make a bulb shine brighter.

The battery's voltage affects the brightness of the bulb, so a stronger battery will lead to a brighter bulb. However, 1.5V cylindrical batteries come in four different sizes! The idea that a bigger battery contains more electricity (because it contains more chemicals) is not necessarily true: it depends on which chemicals, and the quality of them. This is why each size of commercial battery is available in a confusing range of grades: 'long lasting', 'extra power' and so on.

You need a wire from each end of the battery to light a bulb, because there is not enough electricity in one end.

The first part of the statement is right, but the reason is wrong. It often pays to ask *Why?* in order to assess a child's understanding. If the reason stated above were correct, you should be able to connect one end of a battery and the other end of another battery to a bulb and get it to light, without connecting the other ends of the batteries. Of course, this will not work: it is a complete *circuit* that is needed.

Electricity comes out of both ends of the battery and runs together in the bulb.

The 'clashing currents' theory, as this is called, is quite interesting. It contains the idea that a complete circuit is needed, involving both terminals of the battery, as well as the idea that energy is coming from the battery. It shows that the child is attempting to bring together observations. Like all theories that are based on the available evidence, it needs to be tested against further evidence.

A diode, or 'one-way valve', will show that the flow of electricity has a single direction. Unfortunately, introducing either a buzzer or an LED into a circuit in series with a bulb will mean that the LED or buzzer will work, but the bulb won't: the current will be too low to make it light up. Even so, the 'clashing currents' theory suggests that the LED or buzzer should work either way round. I have known one child to argue that an LED only works one way around because *different* types of electricity come out of each terminal – that's why they clash! This theory was disproved by using a terminal from each of two different batteries and observing that there was no complete circuit between them.

Teaching ideas

Live and dead bulbs (testing and sorting)

Give the children a box of 1.5V bulbs and a 1.5V battery, and ask them to check and sort the bulbs. (This always used to save me a lot of time when a class returned the 'electricity box' to the science cupboard!) Extend this by asking the children to use two 1.5V batteries (in a battery box) to test and sort a range of bulbs: 1.5V, 2.5V, 3.5V and 6V. Tell them that they don't need to test each bulb for long: a quick flash is enough. With a bit of practice, they should be able to distinguish quickly between 1.5V (very bright), 2.5V (bright), 3.5V (less bright) and 6V (dim). Note that a 1.5V bulb will 'blow' if it is left connected to a 3V supply for any length of time.

Batteries and bulbs (observing and sorting)

Make a collection of different batteries and bulbs (avoid using 'button' batteries of the wristwatch type with younger children, as they can easily be swallowed). The children should identify the positive and negative battery terminals, then sort the batteries by voltage, by shape and by size. Next, they should identify the terminals on different bulbs and note similarities and differences in size, shape, voltage and filament. They should match batteries to bulbs where possible. **NB** Make sure they are aware of the dangers of handling delicate glassware.

How does your circuit flow? (modelling and explaining)

Demonstrate the skipping rope model (see page 23) to help the children understand the flow of electricity. Invite comments on any problems with the model.

Concept 5: Taking control of electricity

Subject facts

Controlling circuits

Most circuits are simply 'on' or 'off': if the circuit is complete, the bulb (or other component) will work; if the circuit is broken, it won't. A **switch** is a means of simply completing or breaking a circuit without having to take the circuit apart and put it back together each time. The circuit shown in Figure 3 on page 28 is in need of a switch, since the only way to turn it off without unscrewing the bulb holder or undoing pieces of tape would be to unscrew the bulb itself (which might be hot) or to cut the wire.

Simple switch

To make a very simple switch, cut the wire in a circuit and strip the ends. When you touch the bare wires together, the bulb lights up. All switches work on this basic principle: the rest is technology! You can buy switches or make them (see Teaching ideas on page 31).

Variable switch

A simple switch is either on or off. A variable switch, such as the volume control of a radio or TV, allows you to decide how much electricity you will allow to flow through the circuit. In classroom circuits, a variable switch takes the form of a length of high-resistance material that will allow electricity through, but not very well. When placed in a circuit with a battery and bulb, it resists the flow of electricity in a variable way: the greater the length of resistor material the current has to flow through, the dimmer the bulb will be (see Figure 3, page 28). Unless you actually break the circuit, electricity will continue to flow, even if the bulb is so dim that it does not seem lit at all.

Figure 3 Using a higher resistance material as a variable switch

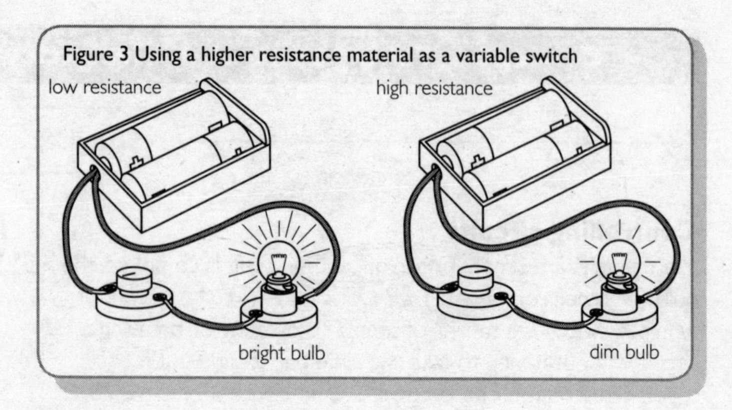

Bulb

If you study a small **bulb** carefully, you will see a very thin piece of wire (the filament) held between two upright prongs. This filament heats up and then glows brightly when electricity passes through it (due to its resistance). The greater the voltage, the brighter the bulb glows, until it burns through and breaks the circuit. This form of electrical light source, the filament lamp, is not the only one that children may be familiar with. In many light sources, electricity is passed through a gas-filled tube to make the gas glow. Light sources of this kind need less electricity to produce the same brightness, but they cost more to make in the first place. Examples are fluorescent tubes (often found in shops and classrooms), street lamps and halogen lights.

Electric motor

When electricity passes along a wire, a magnetic field is generated around the wire. This field can be made stronger either by increasing the amount of electricity passing along the wire or by coiling the wire up. A coiled wire must be coated with an insulator to stop the electricity taking a short cut by jumping from strand to strand where they touch. When two magnets are brought next to each other, they will either attract or repel depending on whether the poles are similar or opposite (see Chapter 2, page 44). If a coil with electricity flowing through it is brought close to a magnet, it will either be repelled or attracted depending on its orientation relative to the magnet.

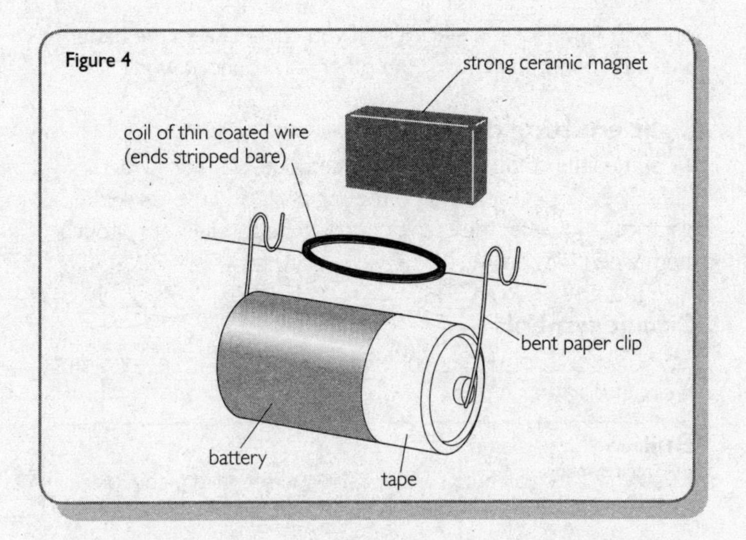

Figure 4

These forces of attraction and repulsion make a **motor** work (see Figure 4). Electricity flows through the coil, causing a magnetic field to form. When a ceramic magnet is brought near, the coil will either push away or pull itself nearer. As the other side of the coil comes around, that will do the opposite – so the coil will pull up, push away, pull again… causing it to spin. This is the principle of the electric motors in such devices as food mixers, vacuum cleaners and milk floats.

The process can be reversed: if, instead of using electricity to make a coil move in a magnetic field, you set up a coil in a magnetic field and then turn it, an electrical current will be *induced* in the coil. This system is used in all electricity-generating devices, from bicycle dynamos (if you turn the pedals, the lights come on) to power stations.

If you connect a small electric motor into a circuit, it will spin in one direction. If you connect the wires on the motor terminals the other way round, the motor will spin in the other direction. Videos providing further instructions for homopolar motors can be found on websites such as Youtube®.

Resource: **www.5min.com/Video/How-to-Make-a-Simple-Homopolar-Motor-5722862**

Buzzer

Most buzzers contain a device called a diode which acts as a 'one-way valve' for the electricity. If you connect it to the battery one way round, it will work; the other way round, it won't.

Light emitting diode (LED)

All of the little coloured lights on a computer, hi-fi or TV are LEDs. They are cheap, have no moving parts and use very little electricity. However, they can get hot, so be careful not to touch them when they are lit.

Circuit symbols

Figure 5 shows some conventional symbols used for components in circuit diagrams.

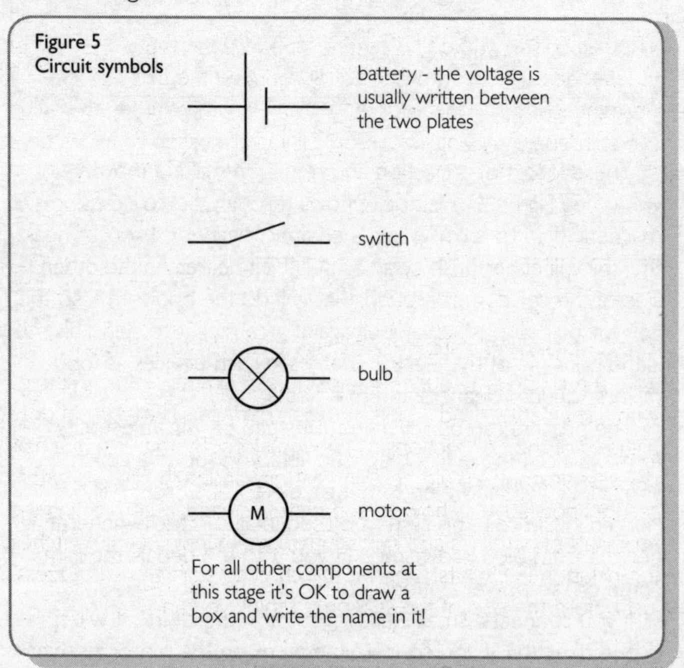

Figure 5
Circuit symbols

Why you need to know these facts

Once an electrical circuit is operating and the current is flowing, the next step is to control the flow using switches. There are a wide range of electrical components which, when connected into a circuit, can do different things. When building circuits, it is important to know the purpose of each component.

Vocabulary

Bulb (lamp) – a device which, when connected into a circuit, will resist the flow of electricity and heat up, producing light.
Motor (electrical) – a device for changing electricity into movement through electromagnetic induction.
Switch – a device for controlling the flow of electricity in a circuit.

Amazing facts

Thomas Edison in the USA and Joseph Swan in the UK, working independently of each other, both produced the first electric filament light bulb (that worked for any length of time) in 1879. Modern 'energy-saving' fluorescent lights are based on work by George Stokes (1852).

Teaching ideas

Circuit progression (testing and observing)

Take the children through the making of a simple circuit, starting with just a battery and a bulb between each pair. They should draw pictures of the connections as they go along. By the end of a half-day session, they should be using battery and bulb holders with secure knowledge of how to complete a circuit by connecting the terminals of batteries and bulbs. Introduce manufactured switches for inclusion in circuits. Only let the children start to use buzzers just before the end of the session – I think you know why!

Make it (design and technology)

Once the children can make simple circuits with switches, challenge them to apply their circuits in practical contexts. For example: light a room in a doll's house; make a door buzzer; make a lighthouse.

Control it 1 (design and technology, problem solving)

Construct a range of different switches (see Figure 6) and ask the children what control situation each would be useful for. Test their ideas together.

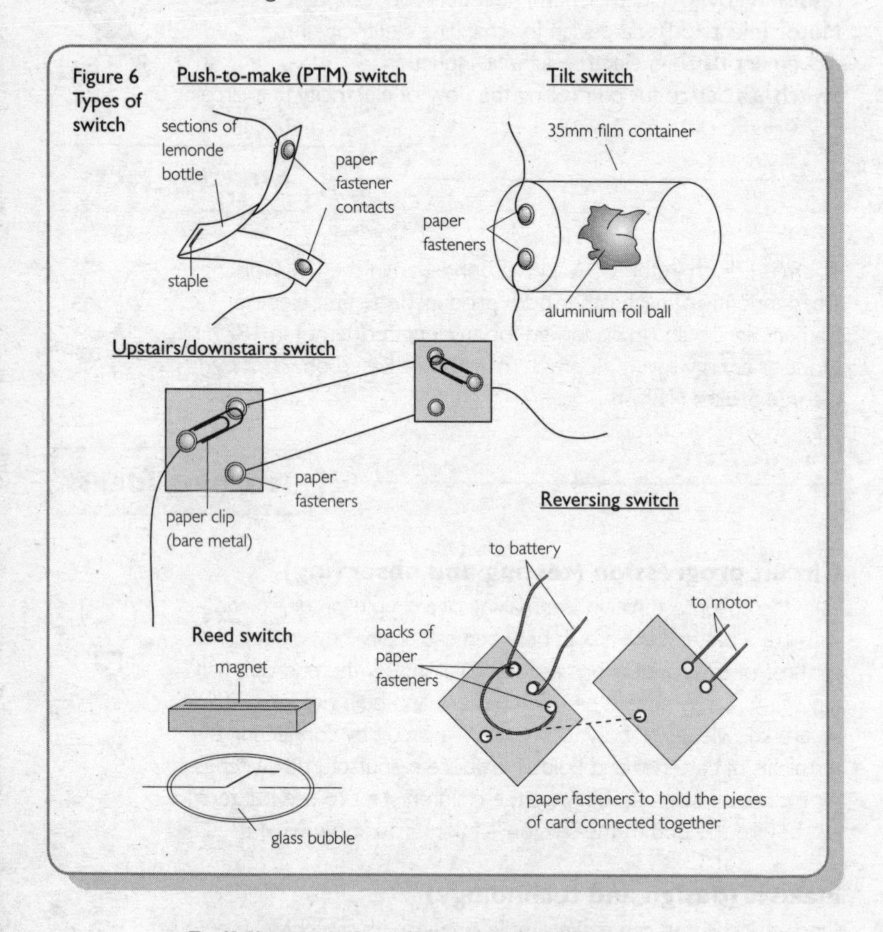

Figure 6 Types of switch

Building to plan (circuit diagrams)

Introduce circuit symbols (see Figure 5). Present the children with circuit diagrams to copy, then ask them to make the circuits. Ask the children to construct circuits and then draw diagrams of them – can other children reproduce the circuits from the diagrams? Give them incorrect circuit diagrams: can they spot the errors? Do they have to build the circuits first?

Control it 2 (design and technology, problem solving)

The children can use a variable resistor in series with a bulb or motor to control brightness or speed. Ask them to suggest how this could be used in a model.

Electromagnet making (construction, investigation)

The children can take a length of wire and coil it around a large nail (starting at one end), then connect the ends of the wire to a battery and see what happens when they take the nail close to a pile of paper clips. Ask: *How could you make the electromagnet stronger?* (More coils of wire, a higher-voltage battery.) *How could you get more coils of wire around the nail?* (Use thinner, coated wire.)

Concept 6: Simple and not so simple circuits

Subject facts

Circuits

There are two basic ways of arranging an electrical circuit: in **series** and in **parallel**.

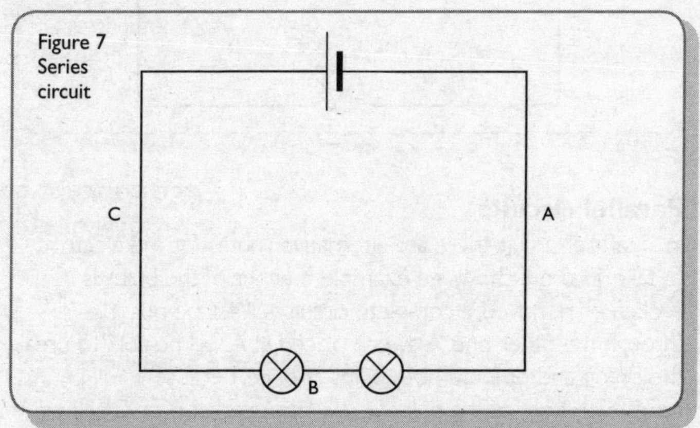

Figure 7 Series circuit

Series circuits

In a series circuit, the electricity has to flow though all of the components in order to complete the circuit. Figure 7 on page 33 shows an example: if either of the bulbs is broken or removed, the circuit will be broken and no electricity will flow. A switch placed at A, B or C will have exactly the same effect: the circuit will be broken and neither bulb will light. The bulbs will each be dimmer than if there was only one bulb in the circuit. In effect, the power coming from the battery has to be 'shared' between the bulbs.

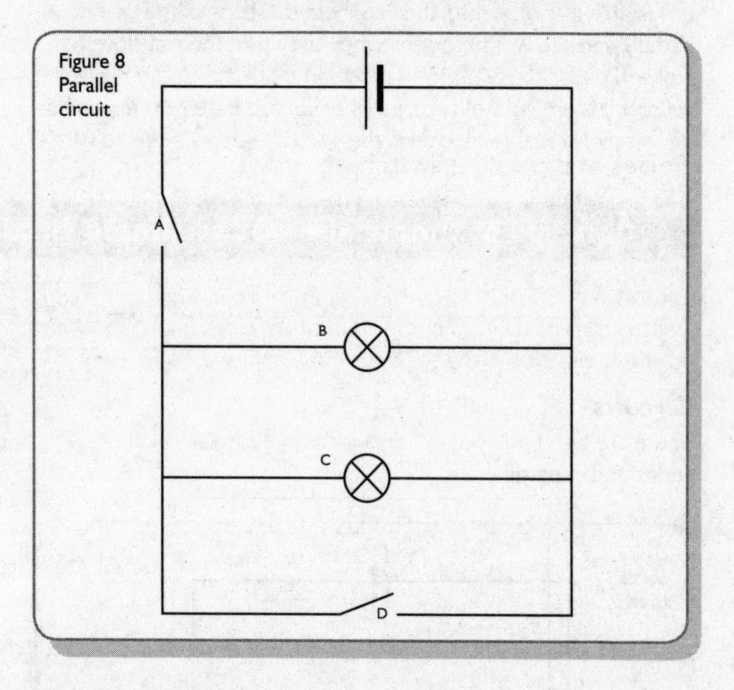

Figure 8
Parallel circuit

Parallel circuits

In a parallel circuit, there are alternative routes for the electricity to take. Figure 8 shows an example: if either of the bulbs is broken or removed, a complete circuit will still be possible through the other one. A switch placed at A will be able to break the circuit and operate both bulbs at once, because it will be in series with both of the bulbs. A switch placed at B or C will only break the part of the circuit going to one particular bulb: it will be in series with only one of the bulbs. For the first bulb to work, switches A and B must be closed; for the second bulb to work, switches A and C must be closed. Note that if the switch at D were closed, both bulbs would go out since there would be a 'short circuit'. Electricity always takes the path of least resistance, and a connecting wire offers much less resistance than a bulb.

If switches A and B are closed, it will be just the same as having one bulb in a simple circuit. However, when switch C is also closed, *both* bulbs will be just as bright as one bulb on its own. The catch is that there is only so much energy stored in the battery: to make both bulbs light up as if they were on their own uses up that energy more quickly. The flow of electricity will be greater, causing more power (watts) to be used. Since the battery's voltage cannot change, this means that the current (amps) is greater.

Series and parallel switching

The arrangement of switches can be used to control a circuit in different ways. Two switches can be arranged in series (as in Figure 9) so that *both* switches need to be closed for the circuit to work. Most domestic wiring is arranged like this, so that one switch acts as a safety check for the other. For example, you have to switch on at both the socket *and* the appliance for a food mixer to work.

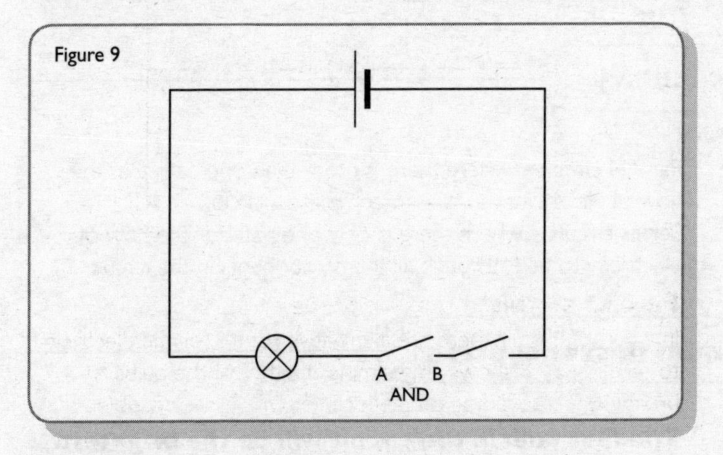

Two switches can also be arranged in parallel (as in Figure 10 on page 36), so that *either* switch can be closed to complete the circuit. For example, an intruder alarm may be triggered by the opening of either the front or the back door.

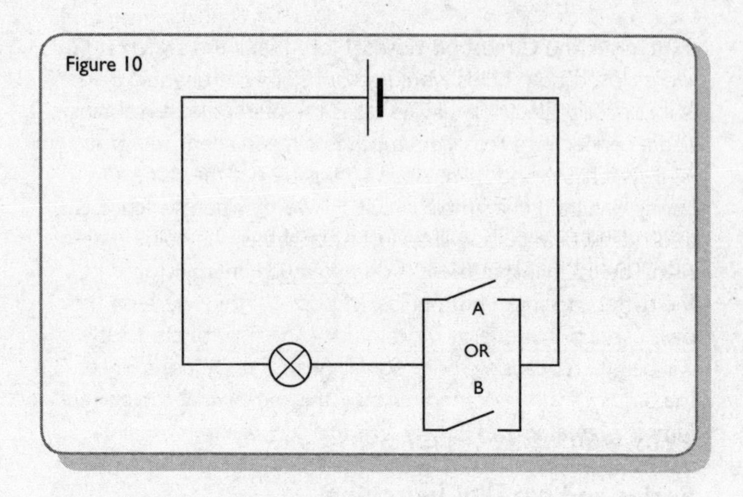

Why you need to know these facts

Components can be connected together to complete a circuit in different ways. These different arrangements affect how the components operate, so an awareness of this is important for constructing and using circuits.

Vocabulary

Parallel circuit – where there is more than one path that the flow of electricity can take to complete a circuit.
Series circuit – where there is only one path for the flow of electricity to take, through all the components in the circuit.

Common misconceptions

The first bulb in the circuit will be the brightest.
This misconception often arises where children have been introduced to the idea that 'electricity comes out of the battery'.

The assumption is that the electricity has to light the first bulb first – then, if there is any electricity left over, it will light the next bulb. Sadly, because the bulbs used in classroom circuits are of poor quality, there is often a marked difference in brightness between two supposedly identical bulbs placed in series. Swap the bulbs over in their bulb holders: the one that was brighter before should still be brighter; so it is the bulb itself, not the position of the bulb, that is the important factor.

Questions

Why won't this circuit work?

This is probably the most common question when children are constructing circuits! Use a methodical approach to help the child check that all the components are in place:

1. Have you got a complete circuit? Follow the path from one terminal of the battery back to the other terminal. If you can't, reconnect the components to make a complete circuit.
2. Is there a short circuit somewhere? Can you complete a circuit without going through any of the devices? (Check especially that wires behind devices aren't touching.) Make sure that the electricity must pass through at least one device in any branch of the circuit.
3. Are the connections secure? Check that the wire has been stripped back enough to make good contact, and that the screw connections are tight (especially bulbs in bulb holders) and show no signs of corrosion.
4. Do the batteries work? Check that each battery used is capable of lighting a bulb of the same voltage.
5. Are you using balanced components (eg a 1.5V bulb with a 1.5V battery)? If not, swap for ones of a suitable rating.
6. Does each of the components work? Check each one, individually, with a battery that you know works.
7. Ask someone else to look at it – they might spot something that you've missed.
8. Take it apart and build it again – carefully this time!

Teaching ideas

Series and parallel (predicting and testing)

Encourage the children to connect additional bulbs or batteries into a simple circuit after predicting what will happen.

- *Does it matter which way round a second battery is connected into the circuit?* (Yes.)
- *Does adding an extra bulb change the first bulb's brightness?* (Yes if in series, no if in parallel.)
- *Does it matter where you put the switch?* (No if in series, yes if in parallel.)

Mend it (problem solving)

Give the children circuits that don't work and ask them to mend them. Ask them to make incomplete circuits for another group to mend.

Resources

The vast range of available resources for the teaching of electricity in the primary school can be confusing. BBC Education offer useful animations to support learning in this area such as **www.bbc.co.uk/schools/scienceclips/ages/10_11/changing_circuits.shtml**. It is possible to build up an effective set of compatible resources if you stick to a few simple rules:

- Keep it simple! Go for simple components rather than packaged sets. The packaging and connectors that come with electricity 'sets' can often confuse and disorientate children.
- Buy components that have the same (or nearly the same) voltage ratings.
- Buy enough components to allow for breakages.
- Batteries don't last for ever! Make sure that there are enough fully charged batteries before you start. Rechargeable batteries are discussed on page 20.
- It is important for the children to see how the components are connected, and how the electricity is able to flow, in any circuit that they make.

The three basic elements of any electricity kit are the *power source*, the *output components* (the bits that do things) and the *connectors*.

Power sources

My preference is for non-rechargeable batteries. Power packs (transformers) that can be plugged into the mains supply to provide a low voltage require you to find several sockets, and to contend with trailing wires. Small batteries range from 1.5V to 9V. A battery tester is also an essential item!

Output components

Your key output component is the bulb or lamp. This will need to match the voltage of the battery as closely as possible. The best matches are:

Bulb	Battery
1.5V	1.5V
2.5V	3V (two 1.5V batteries)
3.5V	three 1.5V batteries in a battery box
6V	6V (or four 1.5V batteries)

Using a 2.5V bulb with a 3V battery (two 1.5V batteries in a battery box) means that you have a nice bright bulb without too much risk of it blowing; and when the battery begins to deteriorate, the bulb continues to glow well for a long time.

You may also need: buzzers (3V); motors (3V); light emitting diodes or LEDs of different colours (3V); ES (Edison screw) bulb holders for the bulbs.

Connectors

Crocodile clips can be vicious. I have known some children to consider them excellent clip-on earrings – but not for their own ears! Even when used properly, they can give little fingers a nasty nip. If you must use them, find ones with the plastic sleeve covering most of the jaw.

Connecting wire can be bought by the bobbin and cut using scissors (not your best ones) or snips. Use wire strippers (the ones with large teeth are best), *not* your own teeth, to bare the ends of the wire. If you are using multi-strand wire, twist the ends together to prevent them getting tangled up.

Switches are very important. You will need to provide several types, in particular:

- Toggle switch – turn it on and it stays on.
- PTM (push-to-make) switch – only on as long as you hold it down.
- Reed switch – either normally on or normally off (the first type being more useful), and operated by the proximity of a magnet.
- Dimmer switch – in effect, a variable resistor which controls the flow of electricity.

Collections of electrically powered toys are also useful, as are collections of materials that the children can test for conductivity. Appropriate electrical components can be collected from educational suppliers such as: **www.tts-group.co.uk**

Freebies!

Electricity companies will often supply schools with free resources to support an electrical safety campaign: booklets, posters and videos. These are intended to help children identify and avoid hazardous procedures and situations, and to demonstrate some consequences of not following safe practices.

Websites

Free electrical safety information can be obtained from:
www.switchedonkids.org.uk/pt_resources.html
http://powerup.ukpowernetworks.co.uk/
www.powerwise.org.uk/
www.bbc.co.uk/schools/scienceclips/ages/10_11/changing_circuits.shtml

Books and CD-ROMs

Scholastic Primary Science: Turn It On!
Scholastic Primary Science: Conserve and Preserve
Scholastic Primary Science: Power Station

Magnetism

Chapter 2

Key concepts

There are three main ideas that need to be covered and understood at the primary level:

1. Magnets have two distinct 'poles'. When like poles are brought together, they repel each other.
2. Magnets exert an attractive force on a small range of metallic materials.
3. When they are allowed to move freely, magnets line up with the magnetic field of the Earth.

Magnetism concept chain

See 'Electricity concept chain' (page 10) for general comments.

KS1

Materials can be sorted into two categories: magnetic and non-magnetic. Magnetic materials are attracted to magnets. All magnetic materials are metals. Magnets can attract or repel other magnets.

KS2

Only a few types of metal are magnetic materials. The Earth has a magnetic field, with magnetic north and south poles. A free-floating magnet will align itself to the Earth's magnetic field. Each magnet has a north-seeking pole and a south-seeking pole. There is an attractive force between unlike poles and a repulsive force between like poles. An object made from a magnetic material can be made into a magnet (magnetised) by stroking it repeatedly with a permanent magnet.

KS3

Magnets can be made out of non-magnetic materials (ceramic magnadur magnets). A compass can be affected by the close proximity of magnetic materials or magnets. Magnetic fields are not disrupted by non-magnetic materials. A flow of electricity through a cable produces an electromagnetic effect. A magnet can be made by placing a rod of magnetic material in an electric coil. A steel electromagnet is permanent. An iron electromagnet is temporary. The strength of an electromagnet is determined by the number of turns in the coil and the current passed through it.

Concept 1: Magnets

Subject facts

The discovery of magnets occurred in ancient times. They were known about and used by the early Greek and Chinese civilisations. The word **magnet** comes from Magnesia, a place in Turkey where the natural magnet lodestone can still be found.

Magnets and magnetic materials

A magnet is an object that is able to exert a strong force of attraction on materials that contain the metals iron, cobalt, steel or nickel. These are called **ferromagnetic** materials. Alternatively, we can describe ferromagnetic materials as materials that are susceptible to the attractive forces of magnets and have the potential to become permanent magnets themselves.

Magnets lose their magnetism permanently when they are heated beyond a specific temperature, the Curie point, because the movement of the individual atoms becomes too great to maintain a static magnetic field. All other materials are either *paramagnetic* (very slightly attracted by a magnetic field) or *diamagnetic* (very slightly repelled by a magnetic field). 'Very slight' here means not detectable by anything but highly sensitive, specialised detectors.

Permanent magnets

Before 1820, the only way to make a permanent magnet was

to stroke a steel or iron rod with a natural magnet (a piece of lodestone). The accepted model of how this works is that a lump of any magnetic material is made up of particles that are themselves 'mini-magnets'. Normally, these magnets will be pointing randomly in different directions, so their effects 'cancel out'. Stroking them with a magnet aligns them so that they all take up the same orientation, turning the magnetic material into a magnet.

Magnets produced in this way need very careful handling, as they are likely to lose their magnetism and revert to simply being magnetic material. Dropping and banging will cause the particles to become random once more as their alignment is disturbed. Materials which can be magnetised (and demagnetised) easily are known as 'soft magnetic materials'. They include nails, paper clips and an iron/silicon alloy called Stalloy, which is used in the core of electromagnets.

Since the discovery of the electromagnetic effect by Hans Christian Oersted in 1820, it has been possible to create a permanent magnet by placing a steel rod in a coil of wire which is then attached to a strong battery. The effect of the electric coil is similar to that of the lodestone (see Chapter 1, page 28). The strength of the electromagnet created depends on the number of turns in the coil and the electrical current passed through the coil.

The strongest type of permanent metallic magnet in widespread current use is an alloy of aluminium, nickel, cobalt, and copper, called alnico. An alternative is an artificial ceramic material called magnadur, which contains powdered ferromagnetic oxides. These and others are collectively known as 'hard magnetic materials'. Although they are permanent as magnets, alnico and magnadur are very brittle and prone to breakage as materials. If you keep dropping a steel magnet, you will end up with a lump of steel; if you drop a magnadur magnet, you will end up lots of tiny magnets. Alnico and magnadur magnets can easily be made into a variety of shapes, including bars and rings, for different purposes.

Most magnets found in toys and used in schools tend not to be of the steel type, because steel magnets are more likely than alloy or ceramic ones to lose their magnetism through rough handling. Often bar and horseshoe magnets, which would have been made from steel in the past, are made from plastic with correctly orientated ceramic magnets inserted in the pole areas.

This tends to make them more powerful and reliable; but it can also mean, particularly in the case of horseshoe magnets, that the magnetic field produced is not what you would get from a continuous steel magnet.

Temporary electromagnets

If a rod of a soft magnetic material (such as iron) is used instead of a steel rod in an electromagnet, the magnet will only last as long as the current is flowing: the rearrangement of the particles in the rod is only temporary. This effect has a number of applications: separating magnetic and non-magnetic materials; lifting and moving magnetic materials; operating bells and switching devices.

Magnetic poles

The poles of a magnet are the regions where the magnetic field is strongest. Magnets tend to orientate themselves in a way that corresponds to the Earth's magnetic field (see page 51 for more details). A magnet can be said to have a **north-seeking pole** (which tends to align itself with the north magnetic pole of the Earth) and a **south-seeking pole**. These are usually referred to as its 'north' and 'south' poles.

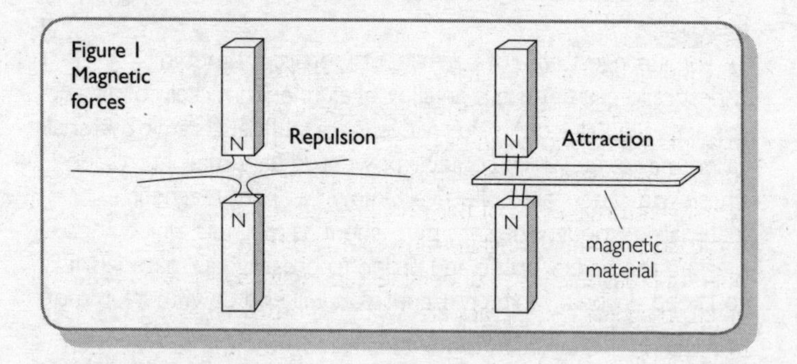

Figure 1 Magnetic forces

Magnetic forces

There is a magnetic force of attraction between a magnet and a magnetic material. However, the attractive force between two magnets is only between opposite poles: the north-seeking pole of one magnet and the south-seeking pole of another. Bringing two north-seeking or two south-seeking poles together will result in a force of repulsion between them. Placing a magnetic material between the two like poles will dissipate the repulsive force acting between them, since both poles will be attracted to the magnetic material (see Figure 1).

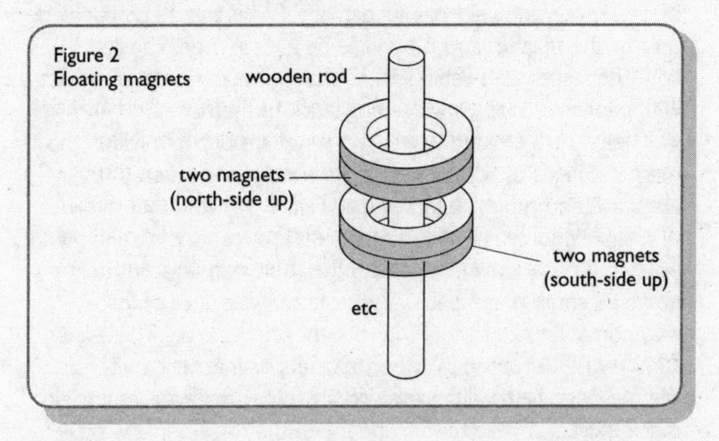

Figure 2
Floating magnets

Alnico or magnadur ring magnets can be placed over a non-magnetic core in an alternating pattern: two north side up, two south-side up, two north-side up and so on. Each pair of rings will 'float' above the one below it, kept up by the repulsive force (see Figure 2). If the top pair are lifted and dropped, they will bounce.

Maglev

Magnetic levitation or 'maglev' can be used as a fairly friction-free means of lifting and moving an object by means of powerful magnets. Some short-run passenger trains, such as those running between passenger terminals at certain airports like Birmingham, make use of this principle, as well as longer, faster tracks such as the Shanghai Transrapid. The train carries a set of powerful magnets on its underside, causing it to levitate when electromagnets in the track are switched on. By altering the orientation and strength of the electromagnets, the maglev train can be made to start or stop, go backwards or forwards and go faster or slower.

Lines of magnetic force

The strength of a magnetic pole reduces with distance from it. Although the force fields produced by magnets cannot be seen directly, their effects on small particles of a magnetic material (such as iron filings) can be.

NB Do not let children handle iron filings directly. The filings stick very effectively to sweaty little fingers, making them difficult to wash off. From that point, if the children rub their eyes, it can cause serious and very painful damage. Filings should be handled only by the teacher, and otherwise be kept in a clear, sealed container (such as a Petri dish). Covering the poles of a magnet with paper will stop filings getting stuck to them.

The pattern created when iron filings are brought within the magnetic fields of adjacent magnets gives an indication of their attractive and repulsive forces (see Figure 3). Note that the lines of magnetic force run straight between adjacent north and south poles, but push away from each other between adjacent north-north or south-south pairs. Also note that the lines of force may come very close together, but they never cross. These lines can only be disrupted by other magnets or magnetic materials: the magnetic field will go through any other materials as if they didn't exist.

Figure 3 Forces shown by iron filings

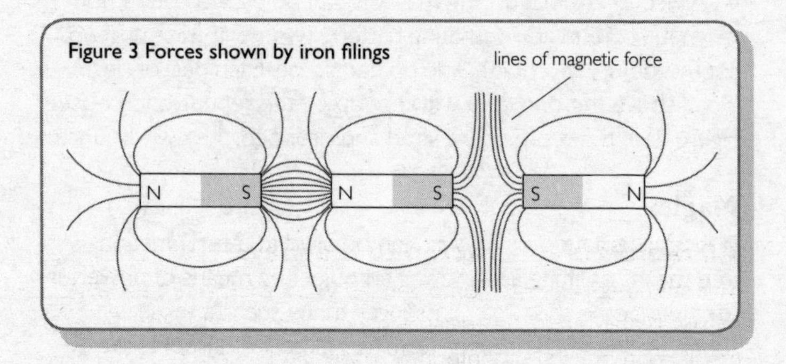

Why you need to know these facts

Most primary school children will have some experience of magnets. They are usually fascinated by the effects of this 'invisible force', which is why magnets often feature in toys and games. The effects of magnets can be observed and described, but there is still considerable scientific controversy over the exact nature of magnetism.

Vocabulary

Ferromagnetic material – a type of metal (iron, steel, cobalt or nickel) that can be strongly magnetised.
Magnet – an object that produces a magnetic field.
Magnetic material – a material that is affected by a magnetic field.
North-seeking pole – the pole of a magnet that aligns itself to the Earth's magnetic north pole.
South-seeking pole – the opposite of a north-seeking pole.

Amazing facts

- The world's largest electromagnet, in Russia, contains more metal than the Eiffel Tower.
- The world's fasted commercial train is the Shanghai Transrapid Maglev which has a maximum operating speed of 268mph (431kmph).

Common misconceptions

All metals are attracted to magnets.

This statement calls for some general work on properties of metals. Metals being called by their colour rather than their actual metal content can be confusing: 'copper' and 'silver' coins are really alloys of several different metals and may not be consistent between currencies (British 'silver' is non-magnetic, French 'silver' is magnetic). 'Tin cans' are not made of tin; 'silver' foil does not contain silver. A set of metal samples from an educational supplier will be a useful investment.

Bigger magnets are more powerful.

This generalisation can be challenged by testing. *What would be a fair test for the strength of a magnet?* (See Teaching ideas below.) Variables to consider might be its mass, what it is made of and its shape. You will find that a small alnico magnet is much stronger than a larger piece of lodestone. The strength of a magnet depends on the extent to which the atoms in the material are aligned.

Questions

What is a magnet?

It is an object with the special property of being able to attract certain 'magnetic' metals. We can only describe a magnet by what it does – there is, as yet, no convincing simple explanation of what a magnet is. The idea that a magnet is made up of 'mini-magnet' particles that are lined up to face the same way may help some older, more imaginative children, but could be confusing to others.

Teaching ideas

Magnetic materials (exploring and sorting)

Give the children a collection of various materials, both magnetic and non-magnetic, and ask them to sort these into the two categories – first predicting, then testing with a magnet. At a later stage, this can be refined so that the children sort samples of different metals.

Exploring magnets (exploration leading to planned investigation)

Give the children a selection of magnets to explore. Once they are familiar with the basic properties of magnets, they can refine their work by setting down clear questions to investigate, such as: *Do bar magnets point north? Are the biggest magnets the strongest? Do magnets work through all materials?*

Concept 2: Magnetic attraction and repulsion

Subject facts

Magnetic materials

The force between ferromagnetic materials and magnets is only one of attraction. This force is not pole-specific: a paper clip will adhere to either pole of a magnet equally well. As magnetic objects adhere to magnets, they become temporary magnets themselves. If you attach the end of one pin to either pole of a magnet, it will dangle vertically; but if you attach two pins to the same pole, the free ends of the pins will repel each other (see Figure 4). This is because they have taken on a *polarity*: if they are attached to the north pole of the magnet, the ends that adhere to the magnet temporarily become south poles, so their other ends become north poles. Similar poles repel, so the free ends tend to move away from each other.

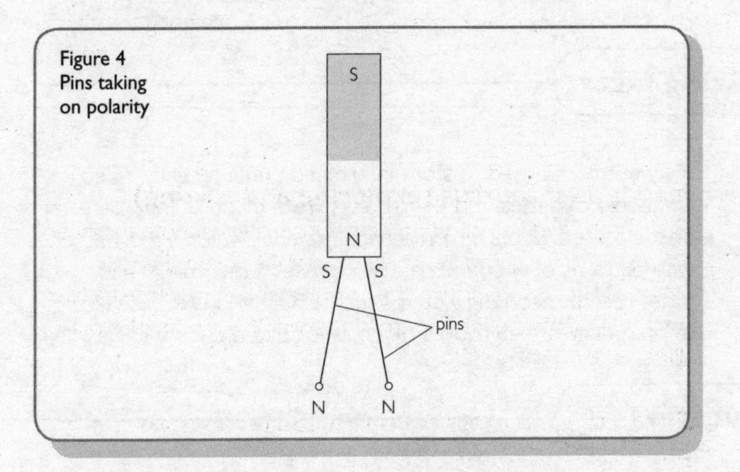

Figure 4
Pins taking on polarity

Magnet blocks

Placing pieces of magnetic material between the north and south poles of a magnet will reduce the magnetic field. When horseshoe magnets are stored, a piece of magnetic material (called a keeper) should be attached to the poles. This has two purposes: it slows down the 'decay' of a steel magnet (its loss of magnetic properties) by protecting the poles from magnetic influences, and it prevents other magnetic objects from being attracted to it.

Why you need to know these facts

An understanding of how magnets interact with each other and with magnetic materials is useful for understanding and appreciating magnet-based toys and games, and other uses of magnets in everyday life.

Vocabulary

Hard magnetic material – can be permanently magnetised.
Soft magnetic material – can be temporarily magnetised.

Amazing facts

Magnetism was used to store information until recently. Video and audio cassettes, and floppy disks were made of thin pieces of plastic coated with a ferromagnetic material, which is magnetised with patterns of encoded data as it passes a recording head. More recent recording systems such as CD and DVD use lasers to read data in the grooves of a plastic-coated disc.

Questions

Why don't magnets stick to all metals?
The honest answer is that nobody really knows. It is something to do with the structure of the different metals, but the exact reason is unclear. You could reply: *I don't know,* or *They just don't,* or *I think it has something to do with the structure of the metal, but I don't think anyone is exactly sure.* The best solution would be to go for the third answer.

Teaching ideas

Strength of magnets (fair testing, observation)
Two simple ways to test the strength of a magnet are:

- *How many paper clips will it pick up?*
- *From what distance will it deflect a compass?*

In the first test, discrete data can be obtained and bar charts produced to compare magnets. In the second test, continuous data will be obtained, making a bar line graph a better option.

Electromagnet making (construction, investigation)

See the activity on page 33 of Chapter 1.

Magnetic pictures (exploration, recording)

Sprinkle some iron filings into a clear plastic container (a Petri dish is ideal for this). Use tape to seal the lid down, so that iron filings cannot fall out and form a safety risk (see note on page 46). The children can work (individually or in pairs) with the iron filings and a magnet. They should shake the container to make an even layer of filings, then bring the magnet up from underneath, and observe and record the patterns that appear in the iron filings. These represent the magnetic field (or lines of magnetic force) produced by the magnet.

Concept 3: Magnetic fields

Subject facts

Earth's magnetic field

The Earth has a core of molten nickel and iron. It is thought that friction, caused by movement within the core, generates electrical pulses. This flow of electricity acts like a giant coil, turning the core into a giant magnet. NB This has nothing to do with gravity: all planets produce a gravitational field, but only those with an iron core seem to produce a magnetic field. Although the Earth's magnetic field has been known about since ancient times, it has still not been adequately explained.

Magnetic poles

The Earth's geographical North Pole and its **magnetic north** pole are in slightly different places, less than 800km apart. The North and South Poles are the ends of the axis about which the Earth rotates. The magnetic north pole is in northern Canada and the **magnetic south** pole on the edge of Antarctica, near Australia. The magnetic north pole was first officially visited in the 1860s –

but if you were to visit that point now, you would be over 100km away from the current magnetic north pole. The magnetic poles are slowly moving, at a fairly constant rate. This is because the magnetism comes from the Earth's core, which is molten and is constantly flowing.

If you were to draw a diagram of the lines of force in the Earth's magnetic field (in the same way as in Figure 3 on page 46), the lines at the magnetic poles would be vertical.

Magnetic compasses

Pieces of lodestone floating freely on water (for example, supported by a thin mat) or dangling on a string have been used to find directions for a very long time. The needle of a **compass**, which is a magnet, aligns itself with the Earth's magnetic field and points to magnetic north. The Earth's magnetic field is relatively weak in any particular location, and is easily disrupted by other magnetic fields.

Concentrations of magnetic materials (such as a metal-framed table or a school building) can lead to an inaccurate compass reading of magnetic north. The flow of electricity (which generates an electromagnetic field) nearby will also deflect a compass.

Testing the Earth's magnetic field

The strength of the Earth's magnetic field can be compared with the strength of a magnet by using a compass. Set up the compass in a place (such as the school playing field) where local magnetic disturbance is minimal, so that the compass reading is accurate. If you then bring the north-seeking pole of a magnet up to the north-seeking end of the compass needle, the needle will turn away to give an opposite reading. Check using the position of the Sun, which will always appear in the southern portion of the sky (in the northern hemisphere) during the middle of the day. (***NB Children should not look directly at the Sun.***) Slowly withdrawing the magnet will reduce the repulsive force until the compass is pointing to somewhere inbetween magnetic north and the pole of the magnet. At this distance, the magnet's field balances that of the Earth (in terms of their effect on the compass).

You can never be entirely sure that the compass is pointing 'true' north. The Earth's magnetic field is not perfect, and there are lots of local variations; but at least this activity demonstrates that the Earth *has* a magnetic field. Carrying out the above activity with the children would be an extension at KS2; it will be enough for the children just to read about the use of a compass, or for you to explain it to them.

Why you need to know these facts

The Earth's magnetic field, with its link to the North and South Poles, is a key area of scientific and geographical study. While geography focuses on the uses of this magnetic field, science examines its properties.

Vocabulary

Magnetic compass – a free-floating magnet which aligns itself with the Earth's magnetic field.
Magnetic north and **magnetic south** – the points on the Earth's surface where the lines of force in the magnetic field are vertical.

Amazing facts

- The Earth's magnetic north pole is moving at an average rate of nearly 2km per year.
- Magnets have been used as an aid to navigation since before written records of travel began. It is not known where in the world magnets were first used for this purpose, but it certainly made finding your way around when you can't look to the sky for direction possible.

Common misconceptions

A compass needle points to the North Pole.
This statement shows an incomplete understanding which may lead to embedded misconceptions later. It is necessary to distinguish between the geographical North Pole and the

magnetic north pole. Children should also be aware that a compass needle can be deflected by local magnetic influences.

Teaching ideas

Compass making (design and technology)

A bar magnet dangled on the end of a string might be expected to settle and align itself with the Earth's magnetic field, but may give inexact results: this field is quite weak, and a slight draft or the winding of the cord may affect the magnet's direction. Another option is to float the magnet on a polystyrene 'boat' or thin cork mat in a bowl of water. The boat will need to be quite small in relation to the bowl, or it will touch the side and affect the magnet's position.

Resources

Magnets

You will need a selection of ceramic or alloy magnets in different shapes (bar, ring, button, horseshoe), plus some steel magnets to compare them with. Steel bar magnets should always be stored in pairs, side by side, with each north pole next to the south pole of the other magnet and with a small piece of steel attached to each end. Steel horseshoe magnets should be stored with a steel 'keeper' going between and covering the poles. Strip magnets (strips of flexible plastic impregnated with magnets) are useful, and can often be attached to dry-wipe boards.

Explanations are available at:

www.explainthatstuff.com/magnetism.html

www.bbc.co.uk/schools/scienceclips/ages/7_8/magnets_springs.shtml

Iron filings:

http://demoroom.physics.ncsu.edu/html/demos/128.html

www.technofrolics.com/products-services/choreographed-iron-dust/gallery/video-to-music/dust-video-to-music.html

Magnetic materials

It is well worth having some samples of different types of metal available for the children to experiment with. These can be obtained with the names of the metals embossed for quick identification. Beyond that, children can use magnets to test objects in the classroom. ***NB Children must not take magnets anywhere near electrical sockets or equipment – especially video monitors, TV sets or computers. Iron filings should only be used when they can safely be enclosed in a transparent container, such as a taped-up Petri dish.***

Figure 5
Compass

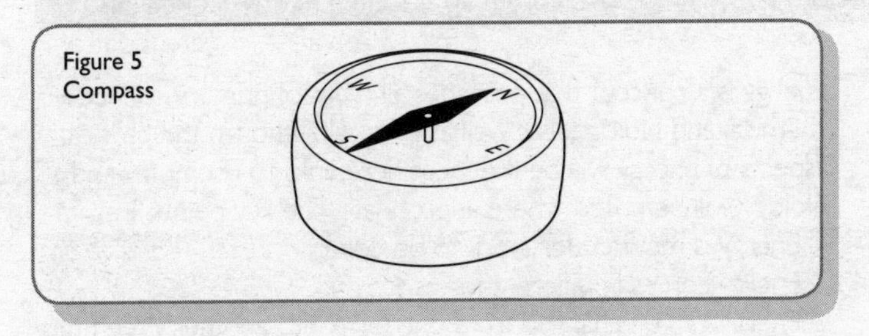

Compasses

Your school should have at least one good quality compass for demonstration, and a number of tiny 'plotting compasses' (see Figure 5) which the children can use to explore magnetic fields.

Books and CD-ROMs

Scholastic Primary Science: Spring into Action
Create & Display: Science and *Create & Display Interactive: Science* (both Scholastic, 2012) have some interesting scientific art activities involving magnets.

Chapter 3

Energy

Key concepts

Energy is a concept that pervades all areas of primary science – chemical and biological as well as physical. Although the physical aspects of energy will be the focus here, links to chemistry and biology will be noted where appropriate. The key points are:
1. Energy is required for work to be done.
2. Energy comes in different forms.
3. Energy can be changed from one form into another. When this happens, some energy is always lost in the form of heat.

Energy concept chain

See 'Electricity concept chain' (page 10) for general comments.

KS1

Introduce energy from a personal, biological perspective: we need energy from food in order to live and do things. In the later stages, begin to introduce physical energy consumption through usage in the home.

KS2

Develop a wider understanding of the flow of biological energy through food chains. Develop the idea that machines also need energy to do things. Begin to introduce the idea that changes in materials involve energy being either used to effect change (as in melting) or released (as in burning).

KS3

Further develop the idea of biological energy flows that start with the energy output of the Sun. Develop the idea of the control of energy use, linked to force and power. Establish the role of energy changes within both changes of state in matter and chemical changes.

Concept 1: Measuring energy

Subject facts

To be capable of doing things – working, running, jumping or even standing still – you need energy. By doing those things, you can be said to be 'using up' energy. We take in energy mainly in the form of food (that is, chemicals from plants and/or animals), and we change it into movement, sound and heat.

For any physical movement or change, energy is needed. This energy is used up by the change or, if already there, released by it. In mechanical terms, the energy used (in joules) is defined as work done – that is, the product of the force (in newtons) used to move a given mass and the distance (in metres) it is moved (see Chapter 4, page 64). For example, lifting a 10kg mass against the pull of Earth's gravity will require 98 newtons of force. If the mass is lifted by 5m, this will require an energy expenditure of 490 joules (the unit of energy). If the same mass is dropped, the same amount of energy is released as it falls.

The more massive an object is, the harder you push it or the further you move it, the more energy you will use. Lifting half of a mass twice as far will use the same amount of energy. The more physical work we do, the more energy we use up – hence the need for food to replace lost energy.

Why you need to know these facts

The concept of energy is often misunderstood and misused within science. It requires careful definition and explanation to distinguish the common usage of the word 'energy' (meaning 'being active') from the scientific concept of energy. Children need to be aware that the concept of energy applies to many different phenomena in a wide range of contexts.

Vocabulary

Energy – the capacity to do work.

Amazing facts

- In the average adult human, 1kg of body fat contains enough stored energy for about 3½ days' worth of normal activity.
- If Britain were to increase the efficiency of its energy use by 1–2%, it would reduce its demand for electric power by 4 gigawatts per year – that's the output of four nuclear power stations as big as the Sizewell plant. A gigawatt is 1000,000,000 watts (joules per second).

Common misconceptions

Energy only belongs to living things.

Well, energy *does* belong to living things. Much of biological science describes and explains how energy flows within and between living things. But understanding energy requires children to go beyond such common usages as 'I'm tired, I haven't got any energy left'. They need to appreciate that many other 'things' need energy: cars need energy from burning petrol; washing machines need the energy of electricity, and so on.

Questions

What are calories?

They are another unit of energy – just as you can measure length in feet or in metres, you can measure energy in joules or calories. One calorie is equal to approximately 4.2 joules. Calories are most often used to measure the energy value of food in terms of its heat output when burned. The Calorie (note the capital 'C') is the unit used in the nutritional information on food packaging, and it is also called the 'kilocalorie' as it is equal to 1000 calories.

Teaching ideas

You should expect and plan to teach general ideas about energy through other areas of science (for example, when teaching about food and nutrition or heating and cooling). It should be a recurring theme when elements of the 'Energy concept chain' (see page 10) arise.

Concept 2: Types of energy

Subject facts

Several different types of energy can be identified at primary level. At a higher level of scientific understanding it is possible to identify just two forms, kinetic and potential; but this theory is best left to secondary level work.

Almost all of our energy requirements can be traced back (either directly or indirectly) to the Sun. By looking at these pathways, it is possible to identify many different forms of energy. Of these, the key ones are:

- Kinetic – movement energy. This could be the energy of a running person, a ball or an **engine**. The movements of air (wind) and water (rivers, waves) are important sources of kinetic energy that can be transformed into other forms, or used directly to move things.
- Light – visible radiation. This is used by plants for photosynthesis. It also enables us to use the sense of vision.
- Thermal – heat energy. No energy transformation is a hundred per cent effective, and some energy is turned into low-level thermal energy at every stage. The idea of thermal energy is important for understanding chemical changes.
- Electrical – the flow of current is a particularly useful form of energy, since it can be transmitted and changed into other forms of energy with relative ease. However, it is relatively difficult to store.
- Nuclear – nuclear fission (the breakdown of atoms) releases intense thermal energy that can be harnessed.
- Potential – 'stored' energy. This has three familiar forms: gravitational, chemical and elastic. Gravitational potential energy

is stored by raising a mass against gravity; it can be released as kinetic energy when the mass is allowed to fall. Chemical potential energy can be released by chemical changes; examples include food (broken down to release energy in the body), plant and fossil fuels such as wood, coal and oil (burned to release heat) and batteries (chemicals react to produce electricity). Elastic potential energy is the 'stretch' in materials (such as springs, rubber bands and rubber balls) which will 'bounce back'.

Why you need to know these facts

Most of the technology that we surround ourselves with is primarily concerned with turning one form of energy into another. Finding more successful, cheap or efficient (energy-saving) ways of doing this is the key area in which technological advances are made. For example, cars have become more efficient in changing chemical into kinetic energy, and thus use less fuel than before.

Vocabulary

Engine – a device for turning stored energy into useful work or movement.

Amazing facts

In every second, an average of 4 gigajoules (4000,000,000 joules) of energy from the Sun falls on each square kilometre of the Earth at the equator.

Teaching ideas

See notes for Concept 1 (page 59).

Concept 3: Using and losing energy

Subject facts

Transformers

We are surrounded by things – both natural and manufactured – that turn one form of energy into another. Plants transform light into chemical energy. Our bodies transform chemical energy from plants (or plant-eating animals) into thermal, kinetic and electrical energy (our nerves and muscles use electrical impulses). We also make machines to transform energy in various ways:

- Car engines change chemical into kinetic energy. Car batteries change chemical to electrical energy, which is converted into light and thermal energy.
- Tumble dryers change electrical into thermal and kinetic energy, and perhaps light energy (the 'on' light).
- TV sets change electrical into light, thermal and kinetic (sound) energy.
- Wind turbines change kinetic into electrical energy.
- Electric torches change chemical into light energy.

Energy losses

Take a rubber ball and drop it onto a hard surface: it will bounce, and will continue to do so for several seconds. The ball starts with potential (gravitational) energy and converts this into kinetic energy as it falls. When the ball hits the ground, the kinetic energy is converted into potential (elastic) energy, which is converted back into kinetic energy as it bounces up, and so the cycle continues.

However, this exchange of potential and kinetic energy does not continue indefinitely. The ball soon stops bouncing and lies still on the ground. Why is this? Firstly, some of the ball's kinetic energy is converted to kinetic (sound) energy in the ground and air each time it bounces. Secondly, the ball moving through the air causes air friction, converting the kinetic energy into thermal energy that warms the air and the ball slightly. Since it is losing energy with each bounce, the ball bounces less high each time.

Anything that moves is subject to friction to a greater or lesser extent, causing some thermal energy to be released. The greater the friction, the greater the heat loss – it has been known for racing car tyres to catch fire!

Why you need to know these facts

Conservation is a key issue in modern life. We are often concerned to conserve living things and their environment; but energy also needs to be conserved, and energy recycling will be crucially important in this century.

Much of our technology is concerned with changing energy from one form into another. Energy never seems to be available in the form that we need: we eat food so that we can move about, keep warm and make sounds; our homes receive electricity so that we can operate machines that perform useful tasks. Whenever the form of energy is changed, some of the energy is wasted as low-level heat that escapes into the surroundings.

All the heat from slightly warm appliances builds up with time, and does not go away. This accumulation of waste energy has had a serious effect on the environment, causing global warming.

Vocabulary

Energy transformer – something that transforms one form of energy into another, as our bodies turn chemical energy from food into movement, heat and sound.
Conservation – the process of reducing our overall energy use and needs by making sure than energy is used more efficiently.

Amazing facts

In each second, the Sun emits 13 million times as much energy as the USA consumes in a year. 3.846 × 10 power 26 watts = 384,600,000,000,000,000,000,000,000 watts. An average sized power station may produce 7000,000,000 watts, so the Sun is the same as 50,000,000,000,000,000 power stations – enough for every person on this planet to have 6000,000 power stations each.

Teaching ideas

See notes for Concept 1 (page 59).

Resources

Some good posters about the uses of electrical energy are available from electricity companies. Machines such as food mixers, gas-powered curling tongs or battery-powered toys can be displayed in school to demonstrate the sources and uses of different forms of energy.

Websites:

www.energysavingtrust.org.uk/Take-action/Money-saving-tips/Energy-saving-tips/Energy-saving-kids
www.solar4schools.co.uk/kids/Games-and-Puzzles
www.childrensuniversity.manchester.ac.uk/interactives/science/energy/renewable.asp
Centre for Alternative Energy:
www.cat.org.uk/
www.sciencemuseum.org.uk/onlinestuff/games/energy_flows.aspx

Books and CD-ROMs

Scholastic Primary Science: Power Station
Scholastic Primary Science: Sound and Light
Scholastic Primary Science: Turn it On!

Chapter 4

Forces

Key concepts

Force is a wide-ranging area of primary science. Children need to develop an understanding of how forces affect both moving and static objects, the effects of gravity and magnetism and the nature of floating and sinking. Specifically, they need to know that:

1. Forces can be used to describe the change in motion or shape of an object (unbalanced forces).
2. Forces can also be seen to affect a situation where no change in the shape or motion of an object is evident (balanced forces).
3. Forces can be measured in terms of direction and strength.
4. The force of friction, which restricts movement, is present in every mechanical action.
5. Gravity is a force of attraction that acts across the distance between any two bodies.
6. Forces interact in situations such as floating and sinking.
7. The forces acting over a given area can be described in terms of pressure.
8. Forces are transferred and applied by machines.

Force concept chain

See 'Electricity concept chain' (page 10) for general comments.

KS1

Forces are pushes and pulls. Forces can make objects start or stop moving. Forces can make objects speed up or slow down. Forces can make things change direction. Forces can change the shape of objects.

KS2

Changes in the shape of objects due to forces can be either temporary or permanent. Elastic materials or objects (such as rubber bands and springs) can push or pull back against forces that are temporarily changing their shape. Gravity and magnetism are both forces that can act at a distance. Gravity is a force of attraction between all objects. The more mass an object has, the stronger its gravitational attraction. Weight is the force that objects exert on the Earth due to gravity. The more mass an object has, the more weight it has. Materials with a lower density than water will float on water. Friction is a force that restricts movement. Friction slows down a moving object. Friction can occur between two surfaces, or between an object and the gas or liquid it is moving through. Forces act in particular directions. The directions and sizes of forces can be shown using arrows on force diagrams.

KS3

Forces acting on an object are balanced when it is not moving. Forces acting on an object are balanced when it is moving but not changing speed or direction. Forces acting on an object are unbalanced when it is slowing down, speeding up or changing direction. The unit of force is the newton. Water pushes up on materials, making less dense materials float. This upward force of water is called upthrust. Objects floating in water will displace a mass of water equal to that of the object. Objects sinking in water will displace a volume of water equal to that of the object. Gravity causes planets to stay in stable orbits around the Sun. The gravity of the Moon causes fluctuations in the level of the sea on Earth. Friction can be reduced through the use of lubrication. Air or water resistance can be reduced by reducing the profile of the object presented to the direction of movement. Machines such as levers, pulleys and gears are used to transmit and apply forces. Machines can be used to magnify forces by applying a smaller force acting over a greater distance. Pressure is the force applied over a particular area. Applying the same force over a smaller area can increase the effect of that force on a surface.

Concept 1: Pushes and pulls

Subject facts

Force

A **force** is a push or a pull. Forces are caused by objects acting on other objects, and can be experienced in a variety of ways. They can vary in strength, direction and duration. In general terms, a force can: start something moving; make something go faster or change direction; make a moving thing slow down or stop; change something's shape (either permanently or temporarily). The action of a force requires the expenditure of energy (see Chapter 3, page 56).

Sir Isaac Newton was the first person to describe force and movement in terms of mathematical relationships and universal 'laws' of motion. Our standard unit of force was named after him: the newton (small n) or N.

Force and movement

Forces and their effects are often misunderstood. This is frequently the result of a misinterpretation of the evidence: the 'full story' has not been taken into consideration, and important elements have been ignored or overlooked. Some of these are discussed under Common misconceptions on page 71.

On the basis of superficial evidence, it would appear that to cause constant movement requires a constant force: if you stop pushing the object, it will stop moving. A closer analysis will reveal that frictional forces act against the motion of a moving object, slowing it down (see Concept 4). If you could get rid of the friction, then once you pushed an object to start it moving it would keep moving with a constant speed and direction. This can only happen in outer space. The assumption that an object slows down after the motive force has been removed needs to be challenged: what is slowing it down? The idea that a frictional force can be balanced by a motive force is discussed under Concept 2 on page 72.

Push a large book across a smooth table, keeping your fingertip in contact with it. When you started to make the book move, you probably felt quite a push against your fingertip; but once it had reached the speed that you wanted it to move at, the pressure on your finger lessened. To get it to that speed, you needed to make the book **accelerate** from a standstill, so you needed to push with a greater force. Once it was moving at a suitable speed, you only needed to push with enough force to overcome the friction between the surface of the book and the table. If you were pushing the book over a rougher surface (with more friction between the book and the table), you would have needed to push it harder to keep it moving at that speed.

Newton's First Law of Motion states that an object will stay as it is unless a net (unbalanced) force is applied to it. Unless there is more force acting on it in one direction than in another, a moving object will keep moving in the same direction at the same speed; and a still object will stay put.

Force and direction 1: getting started

When you kick a football along the ground, what is making it go? Your kick. But as soon as it leaves your foot, it begins to slow down because of friction. Other forces may be involved: the ball may be running uphill or downhill (gravitational force), or be aided or hindered by the wind (air pressure). These forces will either help the ball to keep going or slow it down more quickly. So it is possible for forces to combine and either work together or work in opposition.

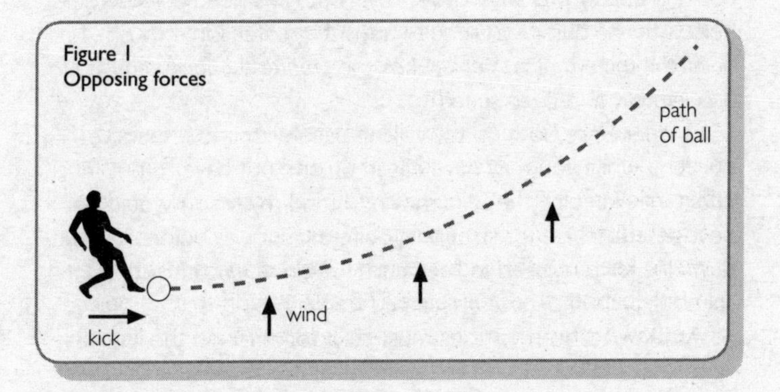

Figure 1
Opposing forces

What happens if the wind direction is at right angles to the direction of the kick? The force of the wind changes the direction of movement, making it veer off to one side (see Figure 1). Note that the ball does not take an L-shaped path: its path curves somewhere in between the directions of the two forces. If the kick were harder, the ball would travel more in that direction; if the wind were stronger, it would travel more in that direction.

Force and direction 2: slowing down and stopping

Sometimes you might apply a force to an object in the opposite direction to its movement. For example, somebody throws you a ball and you catch it. To do that, you have to apply a 'stopping force' to the ball – and the longer the distance it slows down over, the less it will hurt your hands. If you push a loaded shopping trolley to your car, you will probably find yourself desperately pulling back in an attempt to slow it down as you approach the car. In both cases, the force is being applied in the direction of the *change* in the motion.

Again, try pushing a large book across a smooth table. Push the book, then release it and watch it slide across the table. As soon as you have let it go, the only horizontal force acting on it is the frictional force slowing it down, working against the direction of movement.

Put your back into it!

Imagine a road race without petrol, in which the vehicles have to keep pace with each other. A Mini and a double-decker bus are side by side on the start line. I'll let you push the Mini. You could probably push it on your own, but I will need considerable help with the bus: it will require a much greater force (push). *First point: the more mass an object has, the greater the force required to accelerate it to a given speed.*

Once we are both up to walking pace with our respective burdens, while you may have eased up and not have to put much effort into keeping the Mini moving, I (and my team) would be best described as 'not struggling quite as much as before'. *Second point: the force required to keep each vehicle moving at a constant speed is just sufficient to overcome friction.*

At the end, both vehicles must stop together on the line. Both you and I go round to the front of our respective vehicles and apply a force opposite to the direction of movement. Again, you will find that to begin with, it takes a little more effort than expected; but you soon have the Mini slowing gently to a stop. The bus, however, doesn't appear to want to stop. It will take nearly as much effort (friction is on our side this time) to slow it down as it did to get it going. *Third point: the more mass an object has, the greater the force required to decelerate it from a given speed.*

The lesson to be learnt here is a more generalised version of 'If you're given the choice of pushing a bus or a car, always choose the car'. Also, note that two people pushing the bus can get it up to speed twice as quickly as one person can. So the *mass* of the object and the *force* used are both important when you are trying to make an object accelerate (or decelerate). This leads us to Newton's Second Law of Motion: force = mass × acceleration (or $f = ma$). So increasing the force applied to a given mass will increase the rate of acceleration in proportion.

Changing shape

Are you sitting comfortably? If so, can you feel the force that you are exerting on your chair? There are sensors in parts of your body that are currently telling you that you are in contact with something and pressure is being applied. Depending on the nature of the material that you are sitting on, one or both of two things will have happened when you sat on the chair: parts of you will have changed shape and/or parts of the chair will have.

Hopefully these changes of shape are only temporary in nature: shortly after getting up, you and the chair will revert to your original shapes. It is worth mentioning here that two different types of materials are associated with temporary and permanent changes. Elastic materials return to their original shape once the external force is removed. Plastic materials remain in their new shapes.

Stretching and releasing

As was noted on page 65, energy is needed for a force to actually do something. Find a thick elastic band and stretch it out as far as you can. This act requires you to use energy. As you stretch the elastic, you can feel it pulling back at you. You then have to keep up the stretching force or the band will revert to its original shape. If it were a piece of a plastic material such as dough, the initial effort would be all that was needed. But with an elastic material, you need to use your muscles to maintain the tension.

When you release the elastic band, you have two options:

- Release the band slowly, exerting slightly less force than the band, so that it brings your hands back together slowly. This shows that the band is exerting a force throughout the process.

- Just let the band go, and see what happens when that amount of force is applied over a very short period of time.

This situation can also be described in terms of energy. You used energy to stretch the band, so energy has been *stored* in the elastic band. It can either all be released at once (as in a catapult) or slowly (as in a balsa wood plane where an elastic band is twisted and then released to turn the propeller). You also need to use energy to hold a book in the air, because human muscles do not stay taut without expending energy; but a table does not expend energy in keeping a book off the ground.

Unbalanced forces

Throughout this section, it has been clear that where an object is changing speed, shape or direction, the change is caused by a force acting on the object in a particular direction being bigger than the force opposing it. Whenever forces are unbalanced, something will change.

Why you need to know these facts

Force is a phenomenon that we experience constantly – so constantly that we tend to forget about it and ignore it. By observing in greater detail, we can begin to appreciate the effects and characteristics of force. Being aware of forces in everyday life will help children to grasp this important aspect of science.

Vocabulary

Force – a push or pull.
Accelerate/decelerate – to increase or reduce speed of movement in a particular direction.

Amazing facts

The motor racing driver David Purley crashed in 1977, going from 173km/h to 0km/h in 66cm – a deceleration force of 178.9 times the Earth's gravity and survived.

Common misconceptions

Forces are to do with living things.

'A forceful character.' 'May the Force be with you.' 'The life force.' Many common uses of the word 'force' give children the impression that it is something intrinsic to plants and animals, rather than a concept within physical science. As with all scientific words in common usage that have alternative meanings, the key is to model the correct scientific usage for the children and encourage them to use this.

Constant motion needs a constant force.

This is usually true, but only because of friction. Ask the child *why* an object slows down: *Why doesn't it just keep going when you stop pushing it?* Introduce the idea of friction, the force that opposes movement. As long as there is friction in the system, you will need to keep up a small push to overcome it. If you can reduce the friction, you will need to push less to keep the object moving. Pushing a metal block over sandpaper will need more force than pushing it over ice. If you could reduce friction to nothing, you wouldn't need to keep pushing the object to keep it moving.

The same push will always result in the same movement.

Ask the child to push a small box across the carpet, then do the same with a larger and much heavier box. *Is there the same force for the same motion?* Hopefully, the child will realise that mass is a factor too and make progress towards developing an understanding of Newton's Second Law.

Teaching ideas

Exploring forces (exploring, observing)

The children can take a collection of moving toys and explore what happens when they push or pull them, then what happens when they push or pull harder. What happens when they make them heavier and then try to make them move? They can also explore playground equipment: *How do you make the equipment move? What happens when you push harder? What happens when you push in a different direction?*

Pushes and pulls (identifying, explaining)

The children can act out pushes and pulls in PE, both individually and in pairs or groups. Ask them to identify things in the classroom that need pushing or pulling (chairs, doors, pencils and so on), and to explain how these things work.

Squashing and stretching (developing vocabulary)

Place a lump of play dough in the middle of a table. The children can take turns to pick it up and change the shape of it in different ways, saying the appropriate verb (such as 'bend', 'squeeze', 'twist', 'press') as they do so.

Identifying forces (observing, recording)

Children should be asked to look at and think about a range of different situations in terms of pushes and pulls. Good examples would be:

- the playground (swings and slides)
- ball games
- going from the classroom door to your seat.

The children can then talk through various physical actions, using 'pushes' and 'pulls' as the key verbs.

Concept 2: Pushing and pushing back

Subject facts

Balanced forces

Consider the process of stretching an elastic band (see page 69–70). When you were holding the stretched band, your hands were not moving. But you could still feel a force acting on them. Your pull and that of the elastic band were cancelling each other out. In such situations, the forces are said to be 'balanced'.

Forces may also be 'balanced' when something significant is happening – for example, when a skydiver is in 'free fall'. The skydiver is falling at a constant speed: there is no change in the speed or direction of the fall, or in the shape of the skydiver. The forces of gravity and friction (air resistance) acting on the falling skydiver are balanced (see below). When the skydiver reaches the ground, there will be a change in the speed of fall as the forces become **unbalanced** (and perhaps a change in the shape of the skydiver) before the forces are balanced again.

Action and reaction

Consider the book on the table (page 66). Use a bigger book for the following. Place the book on the table: can you see the balanced forces? Gravity is making the book push down against the table; the table is pushing back, stopping the book from falling to the floor. Place your hand between the book and the table: can you feel the book pressing down and the table pushing up? If the book is heavy enough, your hand may be changing shape. These forces are normally acting between the book and the table, but you may not be aware of them unless you intervene in this way.

Now take the book off the table. The book is no longer pushing down on the table, so what is the table doing? Put your arm out straight and hold the book on your upturned hand. Keep the forces balanced: push up as much as the book pushes down. Don't stop pushing up, but take the book away: what happens? Your arm shoots up. Why doesn't the table do this? How does it 'know' when to stop pushing? The explanation is that the table exerts a purely *reactive* force: one that only exists in reaction to another force. This is similar to the way that friction only works to slow something down. When an object stops moving, friction does not throw it back the way it came. Your arm keeps a book in the air by providing an *active* force through the muscles.

All around you, there are examples of balanced forces such as the book placed on the table. Most structures have to resist moving or changing shape when a force is applied to them. If the force is taken away, they will still not move or change shape. Of course, the strength of any structure has limits: if the applied force is too great, it will collapse or move. If an elephant had been placed on the table, the outcome might have been different!

Newton's Third Law: 'It takes two to tango'

According to Newton's Third Law of Motion, for every action, there is an equal and opposite reaction. So any force will be experienced by two things at once. Think of an arm-wrestling match. Who, apart from the two antagonists, really knows whether they are both truly straining or both faking effort? Only those directly involved in the force are able to feel it, in opposite directions.

For example, when you impart a force through a slap on the back, both the hand and the back feel the force. If you immerse yourself in the front row of a rugby scrum, you will feel the force pushing against you – regardless of whether it is your team or the opposition who are doing the pushing.

Terminal velocity

Now imagine yourself as a skydiver jumping from an aeroplane (see pages 72–73). There will come a point in your fall to the ground when you cannot get any faster. Why is that? Surely gravity will have its effect all the way to the ground. As you fall, you will experience an updraft from the air caused by your rapid downwards motion through it. As you move through the air, it pushes back against you. There will come a point when it is pushing back against you as hard as gravity is pulling you. At this point, you will not be able to fall any faster: you will have reached terminal velocity.

As this upwards push of the air, or air resistance, increases in proportion to the surface area you are presenting to the air beneath you, the only way to fall faster is to reduce your downwards surface area by making yourself more 'streamlined'. If you have been falling in the 'belly-down, limbs-out' position, you are clearly not trying to go very fast. Your fastest possible speed in that position will be $55ms^{-1}$ (metres per second) as determined by experiments. A head-first or feet-first dive will allow a much more streamlined (low air resistance) fall, giving a speed up to $80ms^{-1}$.

However, I suspect that when you jump out of an aeroplane, your intention is to make as slow a descent as possible. Gravity is still pulling you down with the same force: you cannot change that. The best strategy is to make yourself as un-aerodynamic (or as poorly streamlined) as possible by increasing your downward-facing surface area. A handy way of doing that is to use a parachute. This makes the upthrust from the air much greater, effectively causing air resistance to balance the downward force of gravity at a much lower terminal velocity. You are still falling, but you will hit the ground at a much lower speed than you would without the parachute.

Why you need to know these facts

Children will constantly encounter opposed forces balancing each other in everyday life. A detailed examination of such situations will allow greater understanding.

Vocabulary

Balanced forces – where the forces acting on an object are equal and opposite, causing no change in the object's speed, direction or shape.
Unbalanced forces – where the forces acting on an object are greater in one direction than in another, causing a change in its speed, direction or shape.

Common misconceptions

If a body is not moving, then there is no force acting upon it.

The child needs to consider situations where forces are balanced and there is no movement. Suggest that he or she pushes against the classroom wall: there will (hopefully) be no movement, but the child should be able to feel a force. Talk the child through the activity of placing a hand between a heavy book and a table to feel the forces (see page 73).

If an object is moving, there must be a force acting in that direction.

If you push a toy car and then let go, it immediately starts to slow down. The only horizontal force acting on it is friction. The child needs to distinguish between movement and *acceleration* (a change in the speed or direction of movement). If there is acceleration, then there is a force in the direction of the acceleration. Whenever something is slowing down, the dominant force acting on it is opposite to the direction of movement.

Teaching ideas

Pushing and pushing back (identification)

Children identify (verbally in a class discussion, or by making diagrams individually or in small groups) situations in which they push and get pushed back (for example: leaning against a wall, sitting on a bench). Help them to see that the push back does not *follow* the original push (except in a fight).

Concept 3: Measuring force

Subject facts

The newton

The SI (International System of Units) unit for the measurement of force is the newton (N). It is defined as the force required to accelerate a 1kg mass by $1ms^{-1}$ in one second. The force of gravity, if unopposed, would accelerate us towards the centre of the Earth at a rate of $9.8ms^{-2}$ – read as '9.8 metres per second, per second' – every second it will go 9.8 metres per second faster than the previous second. Therefore gravity exerts a force of 9.8N for every kg of mass we possess. By multiplying your body mass in kg by 9.8, you can calculate the force of weight (in newtons) that you exert on the ground.

Rounding 9.8 to 10 makes it much easier to calibrate force-measuring machines. For example, a set of bathroom scales can be used as a force meter by multiplying the readings of mass (in kilograms) by 10 to give the force that you exert on the ground in newtons. It works just as well in other directions: if you hold the scales against the wall and push, the reading multiplied by 10 will tell you how hard (in newtons) you are pushing. However, this only works for pushes: to measure pulls, you need a spring balance (see Figure 2).

Spring balances or 'newton meters' have scales that are calibrated in newtons or grams/kilograms, sometimes both on opposite sides. They are only able to measure pulls. For example, you could attach the hook to a door handle to measure the force required to pull the door open; or hang a 1kg mass from the hook to see the force exerted on it by gravity (which should be 9.8N).

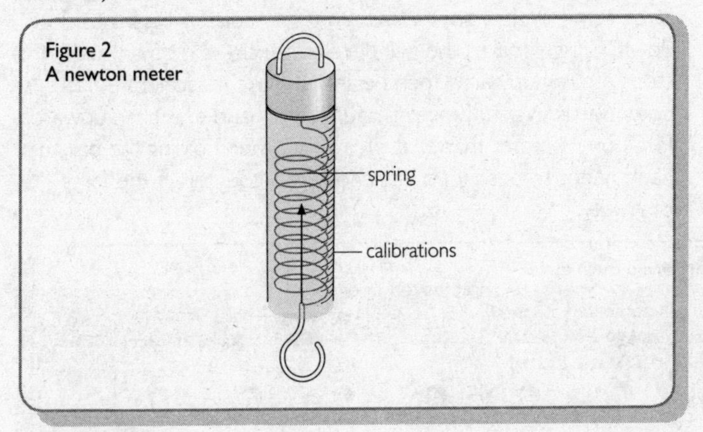

Figure 2
A newton meter

Direction

As well as knowing the strength of a force, you may also need to know the direction that it is acting in.

Consider the example of kicking a football (see page 67). On a windy day, the direction of the wind force will clearly be important. The direction of a force is indicated on a force diagram with an arrow; by convention, the larger arrows indicate stronger forces. Balanced forces should be shown as opposite arrows of equal size; where a change of speed, shape or direction is occurring, one arrow needs to be bigger than the one opposite to it. On Figure 1, on page 67, we could draw arrows to show the forces of the kick, friction and the wind (see Figure 3).

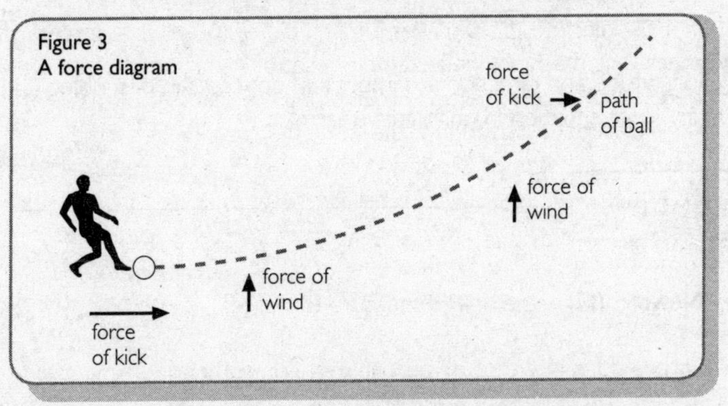

Figure 3
A force diagram

It is easy to confuse the direction of a force with that of movement. Consider someone throwing a ball up and catching it (see Figure 4). As the ball leaves the hand, the upward force of the throw is greater than the downward force of gravity (ignore air resistance). As soon as the ball leaves the hand, the only force acting on it is downwards, so it immediately begins to slow down. Gravity makes the ball decelerate (slows it down) until it stops moving upwards, then begins to make it accelerate back downwards. When the ball is caught, the hand exerts an upwards force on it greater than that of gravity. When holding the ball, the hand needs to push it up just enough to counteract the force of gravity.

Figure 4 Throw and catch forces

Now consider a hand holding a tumbler. The hand has to press against the glass, with the glass pressing back, in order to make sure that the friction (see Concept 4, page 80) between the hand and the glass is sufficient to prevent the downward pull of gravity slipping it out of the hand's grip.

Why you need to know these facts

Magnitude and direction are important concepts in describing any force situation in quantitative terms.

Vocabulary

Newton (N) – the unit of measure for force.

Teaching ideas

The children can make their own force meter using a long, thin rubber band, as follows (see Figure 5). Attach a cut rubber band to the outside of a cotton reel using tape. Thread a length of dowel through the centre of the cotton reel until it meets the rubber band. Attach a 'hook' (bent paper clip) to the band and dowel with more tape. The hook can be used for pulls and the other end of the dowel for pushes.

This force meter can be calibrated for measuring as follows. Holding the reel with the paper clip down, attach a 100g hanging mass to the paper clip. The band will stretch and the dowel will bob down. Where the dowel enters the cotton reel, make a mark – this is how far the band stretches under a force of almost 1N. Now try 200g to measure 2N, and so on. (Be careful not to overstretch the band, or it will slip.)

The children can now measure (very roughly) what force is required to make different objects move or accelerate. They could try to give different toy cars 'fair' pushes, or even give a larger mobile toy a constant 1N push to see how it accelerates. To do this, they should hold the toy still and push the length of dowel into the back of it until the band has stretched to the 1N mark. Then, holding the force meter by the reel and keeping it at the 1N mark, they should release the toy and keep pushing it for as long as possible (if it doesn't move, they should use a smaller toy).

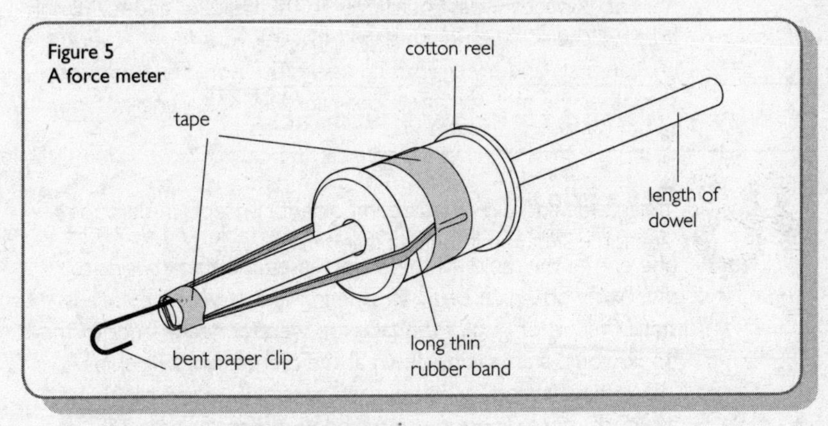

Figure 5
A force meter

Concept 4: Friction forces

Subject facts

The nature of friction

The odd thing about **friction** is that you don't really notice it until you try to move, or try not to. Friction either acts between surfaces or between an object and a medium: it happens when things rub together, or when something moves through something else. Friction is a force that always resists movement. Sometimes this causes problems, and sometimes it actually helps by assisting movement on a different level.

The book and the table, part 3

Between the book and the table, there is the potential for friction, since the cover of the book is in contact with the surface of the table. Slide the book gently across the table: notice how it fails to keep going. How could you increase the friction? Make the surfaces rougher, or use a book with bigger covers (a greater surface area). So to reduce the friction, you would do the reverse: reduce the surface area of the book in contact with the table, or make the surfaces smoother.

To reduce friction, you might also try lubrication: placing some fine solid particles (such as sand) or a slippery liquid between the two surfaces. Press your finger to the tabletop and try to rub it across the surface: not so easy! Now lick your finger (check it is clean first) and try again: is it easier? You should find that the smoother the tabletop, the more effectively the lubricant (your saliva) works.

Get a grip

Now place the book in the middle of the table, and slowly lift one end of the table. As the tilt increases, the book begins to slide. Why doesn't it begin to slide immediately? The reason is static friction, or 'grip'. If the tabletop were perfectly smooth, the book would start sliding down at the merest hint of a slope.

We have already seen how an application of spit can make a surface more slippery – so why do mechanics sometimes spit on their hands to get a better grip of something? It is a matter of quantity. Plenty of water forms a barrier between two surfaces, allowing freedom of movement. A little bit of water only creates surface tension between the water and each surface, thus increasing the grip or suction between the two surfaces. For example, water on the bottom of a glass often causes a mat to stick to the bottom.

A very small amount of water between the book and the table would, in effect, increase the frictional force, so that you would have to tilt the table even more to get the book going. Using a lot of water might not work either: the book might go soggy as it absorbed the water, reducing the amount of water acting as a barrier with the table and creating more surface tension.

Some static friction is always useful: it would be awkward if things always slid off surfaces that weren't perfectly flat. Even the slightest variation in the thickness of the carpet at one end of the room would be enough to send a bowl of fruit on the table crashing to the floor. When you attempt to stand up on ice, there is very little friction between the soles of your shoes and the ice surface (unless the shoes have tiny nails which penetrate the surface). Without friction between your shoes and the surface, you won't get very far except by uncontrollable sliding.

Most means of transport on land depend on creating static friction between the ground surface and the moving surface. You can wear different types of shoe to get better static friction on different surfaces. A sole with a minimum of tread and a maximum of surface area is ideal for smooth surfaces – hence gym shoes and 'slick' tyres on racing cars. When the ground is lubricated with water, these smooth surfaces can become very slippery. The water can form a layer, preventing contact and causing 'aquaplaning' in cars. Normal road tyres and training shoes have grooves to channel the water away, so that contact with the ground can be made. On more uneven and yielding surfaces such as grass or mud, you need shoes with spikes or studs and tyres with coarse, knobbly treads.

Rollers and balls

Instead of using a liquid lubricant between two surfaces, it is sometimes easier to use rollers or ball-bearings. If you want the movement to be only forwards or backwards, placing rollers between the two surfaces will reduce the level of friction. The movement of the rollers is such that the top and bottom are always moving in opposite directions (see Figure 6). If movement in other directions is required, ball-bearings can be used: they can move on a surface in any direction, but the top of the ball is always moving in the opposite direction to the bottom.

Figure 6 A roller on a hard floor

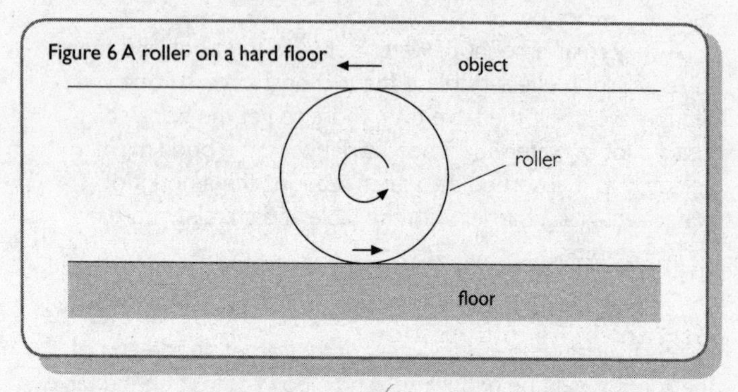

Rollers and ball-bearings are used in many mechanical devices: swivel chairs, bicycle wheels, steam engines, washing machines and so on. It is important to note that this form of friction reduction only works between two hard surfaces that are pushing past each other. If more than the very top and bottom of the rollers or balls comes into contact with either surface, a considerable degree of friction returns. The use of marbles or pieces of dowel under a heavy box will ease it across a tiled floor, but won't be as effective in moving it over a carpeted surface. The balls or rollers will come into greater contact with the carpet as they begin to 'dig in', causing some friction (see Figure 7).

Figure 7 A roller on a soft floor

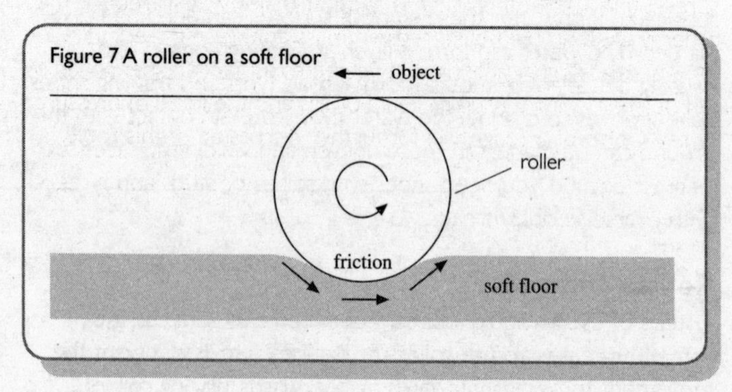

Friction and ice

Why is ice so slippery to walk on? When you have friction between two surfaces, heat is generated. The friction between a shoe or skate and ice underfoot causes some ice to melt – not enough to break though, but just enough for a very thin layer of water to lubricate the surface under the skate or your shoe. As soon as the skate has passed over, the coldness in the ice quickly refreezes the water. A similar thing happens with the runners of sleds, toboggans and skis as they move over snow.

When ice cubes come straight out of the freezer, they feel 'sticky' to handle because they are so cold that they begin to freeze the moisture on your hands. Once they have started to melt a little and have a thin layer of water on them, the surface friction is drastically reduced.

Streamlining and friction

The importance of having a **streamlined** shape when moving through a liquid or a gas has already been mentioned (page 74). The surface area of the object attempting to move through the medium is a key factor. For example, if you are trying to cycle into a strong wind, crouching down over the handlebars will reduce the surface area presented to the wind. Imagine carrying a large piece of board or card on a windy day: would it be easier to carry it face on or sideways?

Keeping the profile presented to the direction of movement as small as possible is one of the key elements in attempting to make anything go faster with the same amount of forward propulsion. Attempting to reduce air or water friction (**drag**) by presenting a smooth surface is another major way to reduce the energy used up by moving a given mass at a given speed.

There are significant links between the resistance experienced by an object moving through a liquid or gas and the force of pressure – see Concept 6 (page 91) for further discussion of this topic.

Why you need to know these facts

Friction is ever-present and can easily be overlooked. Focusing on friction will help children to become more aware of its positive and negative effects in a range of situations.

Vocabulary

Drag – the frictional force experienced by an object moving through a fluid.
Friction – a force between two surfaces that acts against the direction of movement.
Streamlined – a shape which minimises the profile presented by an object in order to minimise the resistance it encounters when moving through a liquid or gas.

Questions

When toy cars come off the end of a ramp, why does the smaller one slow down first?
It takes less force to slow a smaller object (with less mass) than a bigger one (with more mass). So if the same frictional forces are slowing them, the one with the greater mass will go further.

Why don't things stop moving as soon as you stop pushing them?
Because it usually takes some time for frictional forces to slow them down enough to stop them.

Why do some things fall faster than others?
Because of air resistance. The air gets in the way of things that try to move through it, and slows them down – you can walk faster across an empty room than a crowded one where you keep bumping into people. The bigger an object is, the more air it bumps into and the more it is slowed down. So a screwed-up piece of paper will fall faster than a flat sheet.

Teaching ideas

Friction (testing, measuring)
The children can investigate whose shoes have the best grip. They could use a ramp: *At what angle of the ramp does each shoe start*

to slip? Or they could use a force meter to drag each shoe over different surfaces: *Are different shoes better on different surfaces? Which surface would make the best slide?*

Car ramp (investigating, measuring, recording)

The children can carry out a range of investigations using toy cars and a ramp. It is probably best to use standard buggies made using a construction kit (such as Lego® Technic). Variables to explore might include: the angle of the ramp; the surface of the ramp or runoff; the size, number or type of wheels, tyres or axles; the mass or starting position of the buggy; the air resistance (using a piece of card as a 'brake' and varying its size, angle, shape or position); the use of lubrication.

Force diagrams (recording and explaining)

Introduce the idea of using arrows to represent forces. Ask the children to draw force diagrams for some given situations such as kicking a ball, pushing a shopping trolley or opening a door. Encourage them to distinguish between balanced and unbalanced forces.

Concept 5: Gravity

Subject facts

Like magnetism, another force that has an effect at a distance, **gravity** is very poorly understood. We know many things about gravity and its effects, but no one has yet adequately explained why gravity should exist. The force of gravitational attraction operates between all particles of matter; but on the atomic scale, this force is very weak. It is the cumulative nature of gravity – the more matter, the greater the force – which makes it the key force on an astronomical scale.

The existence of a gravitational force which keeps us bound to the Earth was accepted long ago, though this was argued to be the weight of the air keeping things down. It was not until the late 17th century, with the work of Galileo and Newton, that a mathematical understanding of gravity as a force became possible.

Try to think of gravity pulling you towards the centre of the Earth, rather than simply 'down'. When you look at a globe, the idea of 'down' can be misleading: it might be construed as movement across or away from the Earth, rather than towards its centre.

Gravity works between masses, so you attract the Earth with as much force as the Earth attracts you. However, because the Earth's mass is so much greater than yours, its effect on your movement is obvious while your effect on its movement is virtually nil. When masses are particularly large (as in planets, moons and stars), the interaction between them becomes more noteworthy. Strictly speaking, we should acknowledge the gravitational attraction of everything towards everything else; but the effects are usually too small to be observed.

Do heavy things fall faster?

For a long time, it was assumed that heavy things fall faster than light things: that seemed to be 'common sense'. But Galileo's famous 'thought experiment' demonstrated that it could not be true.

Suppose that two steel balls are dropped from a tall building, and the heavier one hits the ground first. What would happen if you tied a cord between the two: would the smaller ball slow down the bigger one, or would the heavier one make the lighter one go faster? Perhaps together they would go faster than the light one on its own, but more slowly than the heavy one on its own? If you made the cord shorter, would that make a difference? Why? If you made it very short, so short that the balls were touching, would that make any difference?

Join the balls together: how fast will they go now? They should go faster than the big ball on its own, if the theory is correct. At what point will the cord be short enough for the pair to go faster than the bigger ball on its own?

The situation doesn't make sense, because the initial assumption is incorrect. It may often be the case that lighter objects fall more slowly: leaves and feathers take a long time and marbles don't. But this isn't a fair comparison, since the effect of the object's shape on the air resistance must also be taken into account.

We really need to test two objects that are identical except for their masses. For example, we could try a cricket ball and a hollow rubber one (the fur on tennis balls might have some friction effect). Drop them together, and they will hit the ground *at the same time*. On one occasion, I caught a child inadvertently dropping the more massive of two balls slightly earlier through a desire to prove that the heavier ball would land first.)

The principle that 'mass doesn't matter' in determining speed of fall can be related to other scenarios, such as a toy car rolling down a ramp. The idea that a heavier toy car will go faster down the ramp than a lighter one, all other things being equal, is an appealing one. But try it! (See Figure 8.) The two trucks will go down the ramp side by side (more or less), and be travelling at the same speed when they hit the bottom of the ramp. At this point, the situation changes. The frictional forces (air resistance and axle friction) slowing both trucks should be the same, but the more massive truck will need a greater force to make it stop within the same distance or a longer distance to stop with the same retarding force. That is why the heavier vehicle will probably go further, but not faster.

Figure 8
Speed/mass investigation

The difference between weight and mass

Mass is the amount of 'stuff' that an object contains. It is measured in kilograms (kg). Denser materials contain more 'stuff' for a given volume than less dense ones. **Weight** is the force that a mass exerts on the ground due to gravity, and is measured in newtons (N). What we refer to as 'weight' in everyday life is really mass. The more mass an object has, the more it will weigh in a given gravity. The Moon's gravity is one sixth that of the Earth. If you visit the Moon, it will not change your mass (you will still contain the same amount of 'stuff'); but your weight (the force that you exert on the Moon's surface) will be only one sixth what it was on the Earth.

If you are floating on water, you will feel 'weightless'. Your mass is unchanged, but your weight is countered by the upthrust of the water. If you were floating and a swimmer swam into you, the swimmer's mass would make a significant difference to the impact: slowing down a mass of 80kg would take four times as much force (for the same change in speed) as slowing down a mass of 20kg. Give yourself a 1N prod with a force meter (see page 79); think of the difference between that and a 4N prod.

In orbit

If you throw a stone horizontally from where you are standing, it will go some distance before being pulled to the ground by gravity. If you throw it harder (with more force), it will go further before it is pulled down. Harder still, and it will go further still.

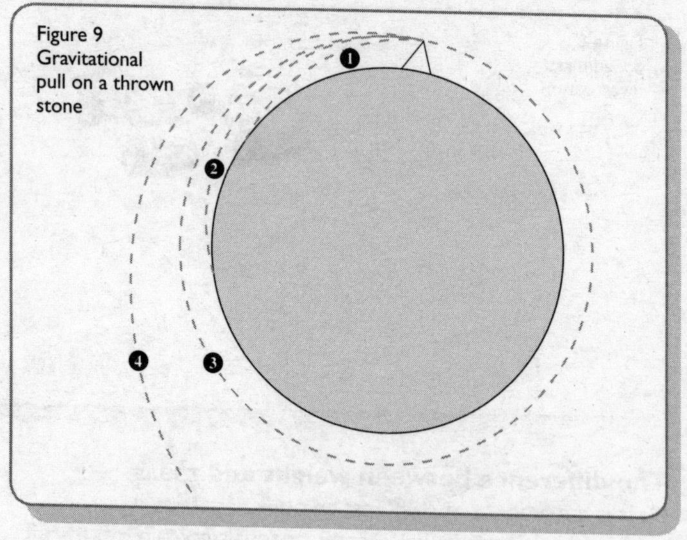

Figure 9 Gravitational pull on a thrown stone

Now consider this 'thought experiment'. If we take these stones to the top of a very tall mountain and start throwing again (see Figure 9), the first throw (1) will go further because we are standing higher up. The second (2), more forceful throw will go further still. With even greater force, the stone may keep falling towards the Earth but, because of the curvature of the planet, keep missing. It will go on falling and missing until it makes it all the way around the Earth. If we ignore the effects of air resistance (3), it should keep going in orbit. An orbit is a continual 'falling and missing'.

In one final effort (4), you throw with even more force. This time, the stone is going further and falling less, so it doesn't go into orbit but spirals out and away from the planet. It has reached 'escape velocity'.

Tides

The gravitational attraction between the Moon and the Earth affects both bodies. Not only are the Moon and Earth held in orbit by their combined gravitational fields about a central point (within the Earth), but the Moon also exerts a significant tug on our oceans (see Figure 10). As the Earth rotates on its axis, the 'ocean bulge' due to the Moon moves around its surface. It does not quite move around every 24 hours, because the Moon is moving as well. On the opposite side of the Earth, there is another 'ocean bulge' caused by the off-centre swing of the Earth/Moon system.

Figure 10 Gravitational attraction between Earth and Moon

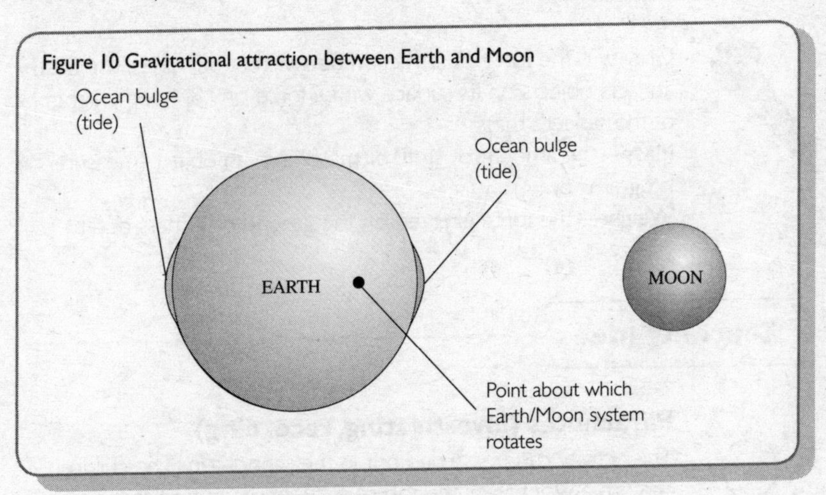

Twice each month, the tides are particularly high, as the gravitational forces of the Moon and the Sun are in line and reinforce each other. Twice each month, the tides are particularly low, as the forces of the Moon and the Sun are at right angles and partly cancel each other out.

Why you need to know these facts

Gravity is not a force that is easy to avoid, and its familiarity can lead children to take it for granted. At primary level, gravity-related activities should focus on appreciation rather than understanding.

Amazing facts

A powerlifter once exerted a force of over 5750N to lift a bar-bell from the ground. The mass of the bar-bell (575kg) was equivalent to almost 300 house bricks – enough to build a wall 2m tall and 2.5m long.

Vocabulary

Gravity – the force of attraction between two masses. The Earth attracts objects on its surface with a force of 9.8N per kilogram of the object's mass.
Mass – the amount of 'stuff' or material in an object (measured in kilograms or kg).
Weight – the force exerted on the ground by a mass due to gravity.

Teaching ideas

Parachutes (investigating, recording)

This activity can be carried out in the school gym. The children can use a wall bar or the top of an 'A' frame as a release point. Ask them: *How much can you slow the descent of a 50g hanging mass?* They could investigate the size and shape of the parachute, the number and position of cords or the type of material used.

Concept 6: Floating and sinking

Subject facts

Density and mass

'Floating and sinking' is frequently regarded as a set of ideas that can be covered through water play early on in a child's formal education; but in reality, it involves using some key scientific and mathematical ideas. One of the first big steps to be made is from 'heaviness' to 'density'. In the same way that children tend to confuse 'volume' with the level of liquid, they often confuse density and mass.

Mass (see page 87) is the amount of 'stuff' in an object. **Density** is the amount of 'stuff' per unit of volume. In general terms, materials that are denser than water will sink in water; those that are less dense will float. The density of water is a building block of the metric system: one kilogram (1kg) is defined as the mass of one litre (1l) of water. One litre is 1000 cubic centimetres (1000cm^3), and 1kg is 1000g, so the mass of 1ml (or 1cm^3) of water is 1g. The density of water is thus 1g per cm^3 (1g/cm^3 or gcm^{-3}). Thus anything with a density of less than 1g/cm^3 will float on water, and anything with a greater density will sink.

Finding out the density

Finding the density of a regular object (such as a cube) is relatively easy. You just measure the external dimensions, calculate the volume, measure the mass (by weighing the object) and divide the mass by the volume. Irregular objects are more of a challenge. Finding the mass poses no additional problems, but how do you determine the volume?

If you submerge an object in water, it will displace exactly its own volume of water. If you fill a bath to the very top, then get in and submerge yourself entirely, you will find yourself mopping exactly your own volume of water from the bathroom floor. If you were to use a bucket with a pouring spout on the side, fill it with water up to the spout and then submerge an object (such as a brick) in the water, the object will displace its own volume in water through the spout. This excess water can be collected in a measuring jug. Now you know the volume, you can calculate the density.

A simple way to compare the densities of an object and water would be to place the object on one side of a balance and the water it has displaced on the other side. If the object goes down, it would sink; if it goes up, it would float.

'Eureka!'

Archimedes' famous cry (meaning 'I have found it!') was his reaction to solving a scientific problem. The king had a gold crown that he suspected was fake. He thought that the goldsmith had stolen some of the gold and added lead, a heavier metal, to make up the correct weight. How could he prove it? Archimedes' solution came to him in the bath. It was to take some lumps of pure gold and balance them against the suspect crown, then immerse both in water to see how much water was displaced. The lumps of gold had a greater volume than the crown, proving that the crown was not pure gold: some of it was a denser metal (lead).

Upthrust

For an object to float on water, the upward push of the water must equal the downward pull of gravity on the object (as with the book and the table). You can see the effect of the upward push, or **upthrust** when an object floats. If you push down on a swimming float in a pool, or a rubber duck in the bath, you can feel the upthrust of the water. You can also feel how much downward force you have to apply, in addition to the pull of gravity, to balance this upthrust. But does this upthrust just 'switch off' when an object sinks?

Think about what happens when you are standing in a swimming pool, up to your neck in water. You can bounce around and almost float in an upright position, as if you were nearly weightless. When you climb out onto the side of the pool, gravity seems to take over once more: it is much harder to lift yourself. So when you are in the water, upthrust or **buoyancy** has an effect on you whether you are floating or not.

Now consider an object that sinks rapidly. You may have a set of blocks of materials at school. Take a metal block, or something similar, and tie some string around it (see Figure 11) so that you can attach a long, thin rubber band. Lift the block by the rubber band and note how much the band stretches. Take the block over to a bucket of water and slowly lower it in. The band contracts as the block is lowered into the water, and reaches a minimum when the block is fully submerged. Note how long the band is now. Take the block out; the band will stretch once more. Put your hand under the block and support it until the band contracts to the length that it had when the block was in the water. You can now feel the force that the water was pushing the block up with: not enough to make it float, but a very noticeable upthrust.

With any object that sinks (that is, is made from a material denser than water), the upthrust will be equal to the weight of a volume of water equal to the volume of the object. So if you held up the block to imitate the upthrust of water in one hand and held the overflow from the bucket (an amount of water with the same volume as the block) in the other, the force on both hands would feel the same.

Figure 11

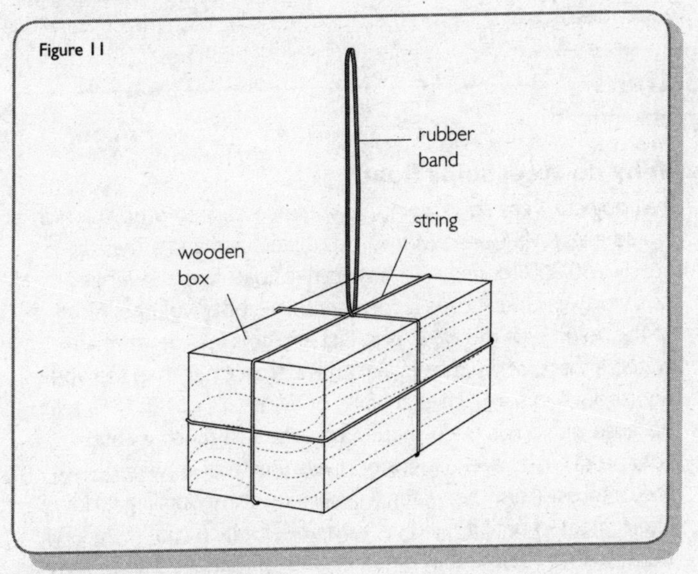

That sinking feeling

You may have noticed that things fall through the air much more quickly than they fall through water. This is partly because water is 'thicker' than air and so it resists movement through it (by friction) more effectively. It is also because the upthrust of water is much greater than that of air: the water pushes up on the object, and may even keep it afloat. For a sinking object, the force of gravity is effectively reduced by the upthrust acting on it.

Why you need to know these facts

From their earliest experience of water play, children will begin to appreciate buoyancy. Such activities can be focused to develop an understanding of the forces involved.

Vocabulary

Density – the mass of an object per unit of its volume.
Upthrust or **buoyancy** – the upward force exerted on a body by a fluid that surrounds it, equal and opposite to the weight of the water displaced.

Questions

Why do steel ships float?

If all objects were solid blocks, it would be easy to predict which would sink. *RMS Queen Mary 2* has a steel hull and a mass of nearly 150,000 tonnes – so how can it float? The answer lies in its shape, which encloses a considerable 'empty' volume. Most of the volume of the *QM2* is in fact air. Air is less dense than water – that's why we have sky above sea, rather than sea above sky (though in some coastal parts of Wales, it's not easy to tell). Because a ship has been shaped to enclose a much greater volume of air, its overall density is less than that of water – so the upthrust from the water will balance gravity pulling it down. *Titanic* floated until it filled up with water, which took its overall density down below that of water and so made it sink. *Titanic* had a mass of only 46,328 tonnes.

Teaching ideas

Plasticine® boats (investigating, recording)

This simple experiment shows the children how to use a material that is denser than water to make a boat that floats. First, they should take a large lump of Plasticine® and place it in water in a displacement bucket (see page 91). The Plasticine® will sink, showing them that it is denser than water and allowing them to measure its volume (the volume of water displaced).

Now they should make it into a 'boat shape'. If they carefully place it on the surface of the water so that it floats, it will displace more water than when it sank. In fact, it will displace a mass of water equal to the mass of the Plasticine®. The children can test this by placing the Plasticine® on one side of a balance and the displaced water on the other.

They might even be able to add some cargo to the boat in the form of marbles. As they add each extra marble, the equivalent mass of water will be displaced – until the boat finally starts to take on water, at which point it will sink. When it does so, the water level will drop, because the Plasticine® and marbles are no longer displacing their own mass of water, only their own volume of water.

Will it float? (testing, sorting, observing and recording)

Give the children a selection of materials (not boat-shaped). Ask them to predict which will float, then test. Can they make those that sank float by changing their shapes?

Bobbing for fruit (sorting and testing)

Give the children a selection of fruit. Ask them to predict which will float and which will sink, then test. Can they suggest which fruit would be best to bob for? Encourage them to try.

Make it sink (exploring)

The children can take materials or objects that float well and try to sink them by hand or by adding weights. How much force does it take?

Displacement (predicting and testing)

The children can use a displacement bucket to measure the displacement of water by objects that float. They should balance the water displaced against the mass of the object.

Boat building (testing and measuring)

The children can make boats out of Plasticine® and float them in a displacement bucket. How many marbles will each boat hold before it sinks? How much water has each boat displaced? Encourage them to compare the mass of each boat (and cargo) with that of the water displaced.

Concept 7: Pressure

Subject facts

Poking and prodding

You're at a party bopping the night away, when another partygoer (a little 'tired and emotional') lumbers over and treads on your feet…twice. The first time it's a stockinged foot, but the second time it's a stiletto heel. Why does the second stomp hurt so much more than the first? The person's weight – that is, the force that he or she exerts on your foot – won't have changed, so why the difference? The answer is **pressure**: the amount of force applied per unit area of a surface. The effect on the surface often depends on the pressure over the area of contact.

Snowshoes are wide, flat shoes for walking over snow. Snow is soft and easily penetrated; spreading the force (your weight) over a larger area makes it less easy to penetrate the snow. Walking over a damp lawn is difficult in stiletto heels, as the heels will penetrate the surface.

Sometimes it is necessary to concentrate a force into a small area. Poking a thread through a piece of fabric with your finger rather than a needle would be impossible. Nails are sharp at one end and flat at the other so that the force of the hammer can be concentrated to a point. If a force is spread over a wider area, you will need more force to do the same job.

Push and push back

When you inflate a balloon, you are increasing the pressure of the air inside the balloon so that it pushes outwards, taking up more space. The air inside the balloon would go on expanding until its pressure was equal to that of the atmosphere, were it not for the stretched balloon exerting an inward pressure on the trapped air. If you let the balloon go without tying the end, it will be able to contract and push the air out rapidly. The reaction to this air movement (see page 73) will mean that the balloon takes off in the opposite direction.

Why you need to know these facts

The concept of pressure will help children to understand how a little force can have a major effect if it is applied over a very small area, such as the blade of a skate or the point of a nail.

Vocabulary

Pressure – force applied per unit area.

Teaching ideas

Poking sand (exploring, investigating)
A sand tray is useful for this. Ask the children to poke a finger into the sand. Now ask them to try with progressively larger squares of thick, rigid card or plywood. They should discover how much more difficult it is to penetrate the sand when spreading the force over a larger surface area.

Concept 8: Machines

Subject facts

Machines are made up from mechanisms, which are ways of turning a small force into a big force or a small movement into a big movement (but not both at the same time). By using **levers**, **pulleys** or **gears**, you can make a small motor move a big load or move a small load more quickly (again, not both at the same time).

Levers

A lever is a simple mechanism. It can be used to apply a small force over a long distance (with the same effect as applying a large force over a short distance). A door is a lever. Try placing your finger on a point near the handle and pushing it open at a steady pace. Now try placing your finger near the hinge and opening the door. You will notice that the force required was much smaller the first time, but that you had to push for much longer.

Here's another example. Think of a see-saw shared by yourself and a small child. How do you get a balance? Since you will be providing the bigger force (weight), you will need to position yourself a shorter distance from the pivot point, or fulcrum, than the child. If you are twice as heavy as the child, then the child will need to be twice as far from the pivot point as you in order to balance you.

Levers are literally all around us, from tools like scissors and hammers to switches like twist knobs and car indicator stalks. All of them work in the same way: the further you get from the pivot point, the less force and the more movement you need to do the same work. Archimedes said that if he had a big enough lever, he could move the world!

Pulleys

A pulley is another way of transmitting and magnifying force by increasing the distance over which a force is applied. Using one pulley (see Figure 12), you need to pull with a force slightly greater than the weight of the load to get it moving, and you need to pull over the same distance that you want to raise the load. However, it is more convenient and easier to pull downwards (working with gravity) than to pull up.

With two pulleys, you only need half of the force; but you will need to pull twice as far to raise the load by the same amount. With three pulleys, it will be one third of the force but three times the distance. The more pulleys you use in practice, the more friction will build up, so you will need to pull with a greater force than you might have expected.

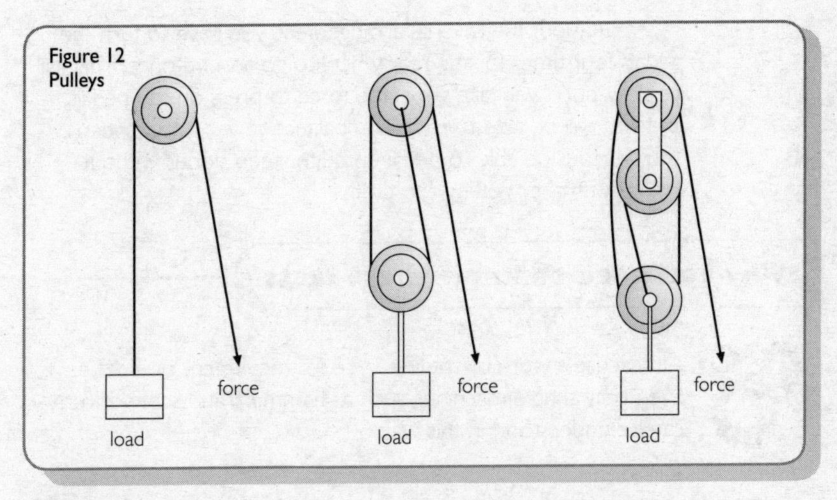

Figure 12
Pulleys

Gears

These work to the same principle as levers and pulleys: scaling force down and distance up, or vice versa. In Figure 13, for every revolution of the big gear wheel (28 teeth), the small gear wheel (seven teeth) will go round four times. If these gears were on a bicycle, linked by a chain, which gear would you attach to the pedals and which to the wheel? If you attached the pedals to the big gear and the wheel to the small gear, then the wheel would go round four times for every time that you turned the pedals. That would be fast; but your feet would have to push with quite a force. As long as you were going downhill, it would be okay.

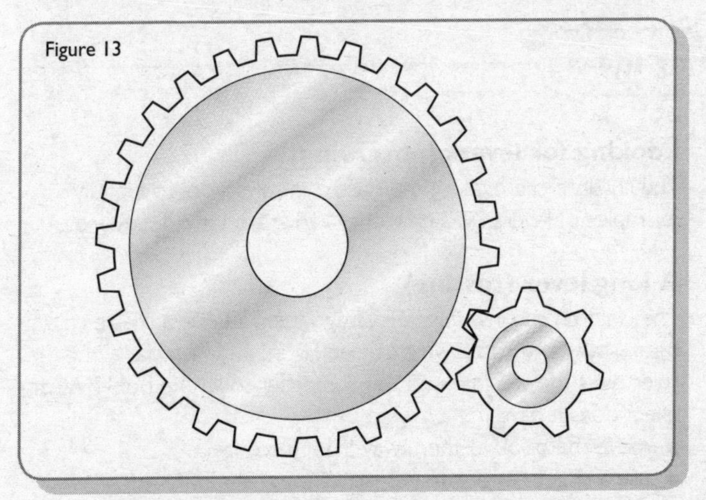
Figure 13

What about the reverse arrangement: you have to turn the pedals four times for the rear wheel to go around once? This will be slow, but it will take very little force to press on the pedals, and you will be able to remain in balance at very slow speeds. You will thus be able to pedal up quite steep slopes without losing control or getting too tired.

Why you need to know these facts

All machines work by trading force for movement or vice versa. Even very simple machines, such as hammers and screwdrivers, can be understood in this way.

Vocabulary

Gear – two wheels with serrated or notched rims that mesh together to transfer movement.
Lever – usually a rigid bar with a pivot point close to one end, allowing a large movement at one end of the lever to be converted into a smaller movement at the other, which effectively magnifies the force applied.
Pulley – a wheel with a grooved rim that allows the transfer of movement via a belt or band.

Teaching ideas

Looking for levers (observing)

The children can explore the classroom or school to identify examples of levers. As a class, they can share their discoveries.

A long lever (testing)

The children can investigate leverage using a table, a metre rule, a book, string and some weights (see Figure 14). How far along the ruler does the weight need to go in order to lift the book? What effect does it have if they:

- move the book further away from the edge?
- use a larger hanging mass?

Figure 14 Investigating leverage

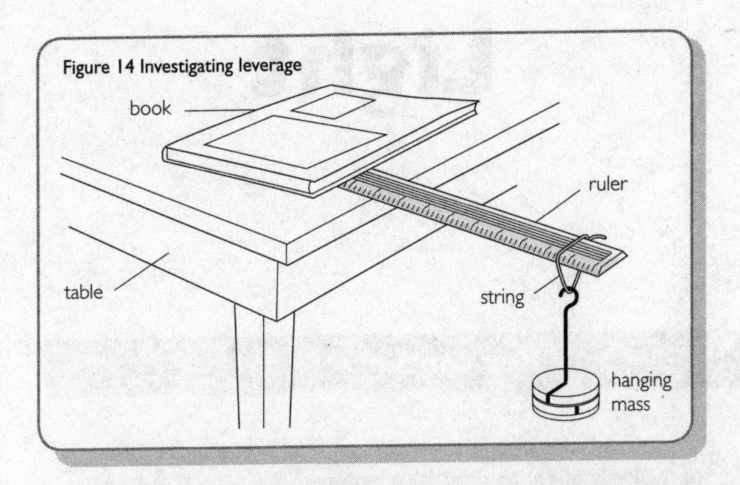

Resources

Provide lots of toys that move (not just cars) for the children to explore. Suitable force measuring instruments should include newton meters, bathroom scales and balances. Floating and sinking equipment should include a displacement bucket and capacity measures. The children should also be able to use outdoor play equipment such as slides, swings and roundabouts to experience the effects of forces.

Websites:

Suitable websites include one on the science of cartoons (links from **www.ase.org.uk/physics.html**), which examines such questions as: *Does gravity only come into effect when you look down and realise that you have overshot the edge of the cliff?* A useful source of ideas is the *Science Zone* website at **www.bbc.co.uk/science**. Skydiving videos can be found on websites such as Youtube®.

Books and CD-ROMs

Scholastic Primary Science: Spring into Action
Scholastic Primary Science: Move It!
Scholastic Primary Science: Lighten Your Load
Scholastic Primary Science: Force Factor
Investigate: Forces (Scholastic) – exciting non-fiction readers

Chapter 5

Light

Key concepts

There are four major ideas in the study of light that are of particular interest to primary scientists:

1. Light comes from a source and travels in straight lines.
2. When it encounters a new material or object, light may be transmitted, scattered, absorbed, reflected or refracted.
3. Light can vary in brightness and frequency (colour).
4. Light can be detected by our eyes.

Light concept chain

See 'Electricity concept chain' (page 10) for general comments.

KS1

Things that produce light are called light sources. The Sun is one of many light sources. We see things with our eyes. We see things either because they produce light or because light bounces off them. You can see things through transparent materials. Light passes through translucent materials, but you cannot see through them. Darkness is the absence of light. Opaque materials block light.

KS2

Light travels in a straight line from a source. Light sources can be seen when light from them enters the eye. Where light is blocked, shadows are formed. Shadows are the shape of the object causing them. Light bounces off (is scattered from) objects, allowing them to be seen. Light travels in straight lines from a source to an object to the eye. More light is scattered from bright objects than from dark ones. Very bright, shiny surfaces allow images to be reflected. White light can be separated into a spectrum using a prism. Red, blue and green are the primary colours of light, and can be combined to produce white light.

KS3

Where light is not scattered by an opaque object, it is absorbed. Human sight is based on the ability to see red, blue and green light. The colour of an object depends on the colours of light that it absorbs and scatters. Light travels at 299,792,458ms^{-1} (metres per second); nothing can travel faster. The path that light takes can be bent (refracted) when it passes from one transparent material to another. Transparent materials can be shaped into lenses and prisms to alter the path of light by refraction. Curving the face of a mirror will alter the reflection. Concave mirrors and convex lenses can magnify images. Convex mirrors and concave lenses can reduce images.

Concept 1: Light sources

Subject facts

Light sources

Light isn't just 'all around us'. It comes from somewhere and enables us to see things. There are only two means by which you can see something: either it is a source of light or it reflects or scatters light produced by a source.

On a dull, cloudy day, it can sometimes be easy to forget that the Sun is our main source of light. The light from the Sun penetrates the translucent cloud cover to illuminate the ground below. Sunlight is also accompanied by heat radiation from the Sun, which is similar to visible light but is beyond our ability to see. When or where there is no strong source of natural light (in the night or in a place that sunlight cannot reach), we make other light sources – usually by burning substances (such as gas or wood) or by passing electricity through them (fluorescent tubes, light bulbs and so on).

In addition to these, some interesting light sources come under the heading of bioluminescence: light from living things.

Examples include certain types of fungi, insects (such as fireflies), and deep-sea fish. These are able to glow by means of a biochemical reaction.

Some highly reflective objects may give the impression that they are **luminous** – but they are just able to focus and direct the light falling onto them effectively. The rear reflectors on bicycles are good examples of this.

Light travels in straight lines

Anyone who has read Stephen Hawking's *A Brief History of Time* will know that light curves when close to a black hole. But as the vast majority of us are unlikely to encounter this phenomenon, the 'straight line' idea is one that we can trust for practical and teaching purposes. Light travels in a straight line from a source; when it changes direction, as it may do after making contact with an object, it always goes off in a straight line.

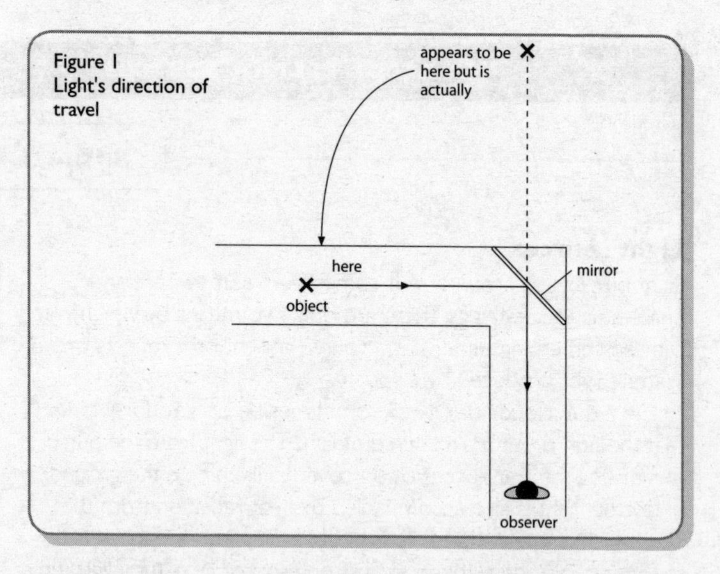

Figure 1
Light's direction of travel

Mirror mazes give us a good idea of how difficult life would be if light did not always travel in straight lines and so we couldn't rely on our eyes to judge the positions of things. In these mazes, mirrors are positioned to give false impressions – for example, that there is a long corridor ahead when, in reality, a mirror has been positioned to reflect the image of a corridor. Someone standing in the corridor would appear to be in front of you when he was actually standing around the corner. You would walk towards him and end up bumping into the mirror (see Figure 1).

If light travelled in curves, you would have similar problems: objects would appear to be in one place, but actually be in another. The ability to see around corners might sound useful – but not if you can't be sure that what seems to be there is actually there!

Another way of proving that light travels in straight lines is to look at the shadow cast by a clear-edged object from a single, small light source. Such a shadow always has clear edges. If light did curve, the edges of the shadow would be blurred because of light spilling around the edge of the object.

The nature of light

Where there isn't any light, there is **darkness**. Darkness isn't a tangible 'thing', it's just the absence of any light source to stimulate our eyes. This can be quite difficult to appreciate, because total darkness is quite difficult to achieve: even at night, there will be stars or street lamps to provide background light. For this reason, our society has become used to having to do something actively to keep light out: curtains, blinds and so on. The only time in my life that I have been in total darkness was when I turned out my helmet lamp when I was in a deep pothole.

Nobody is exactly sure what light is. Sometimes it behaves like a form of wave energy, and at other times it behaves like subatomic particles (which are called 'photons'). Neither theory fully explains what light can do, so the two theories are used to explain different properties of light! This may be a bit confusing and unsatisfying, but scientific models are often not 'realistic': it is just a matter of using the model that works best to explain things.

We know that light (unlike sound) does not need anything to travel through. It is able to travel through outer space (since we can see the Sun and stars), which is mostly empty. This suggests that light is a particle rather than a wave, since waves could be said to need a medium to travel through. However, different colours of light are usually described as having different wavelengths, which implies a wave theory of light. More recently, the colours of light have begun to be understood in terms of particles at different energy levels which are detected by our eyes as different colours.

Another key thing about light is that it travels at a particular speed called 'the speed of light'. Light is the fastest thing there is: nothing can travel faster (apart from spaceships in science fiction). Light travels at approximately 300,000,000 metres per second (see Amazing facts below). However, light only travels at that speed in space: travelling through transparent materials such as air, water or glass will slow it down to varying degrees.

Safety considerations

Very bright light can damage eyesight by 'overloading' and, in effect, burning out the light-sensitive cells in our eyes. Sources of bright light may also be sources of heat (as in a light bulb), and should therefore be handled with care.

Why you need to know these facts

For children to develop a correct understanding of other ideas about light, they need to appreciate that light must come from somewhere, and that it always travels in straight lines.

Vocabulary

Darkness – the absence of light.
Luminous – able to produce light.

Amazing facts

- The speed of light is 299,792,458 metres per second (ms^{-1}). A sprinter can run at $10ms^{-1}$. A car at 70mph is moving at $31ms^{-1}$. The speed of sound in air at sea level is $343ms^{-1}$. Concorde flew at $640ms^{-1}$. Light would travel almost seven and half times around the Earth in one second, if you could make it curve! It takes almost 8 minutes and 15 seconds for the Sun's light to reach the Earth.
- The world's largest solar power plant in Andalusia, Spain, uses 600,000 mirrors to focus the heat of the Sun onto a boiler.
- A solar-powered car crossed Australia in 33 hours in 1996 – an average speed of 85 kilometres per hour.

Common misconceptions

The Moon is a light source.

Reflective objects in space at night can be confusing to us, because they appear to shine in darkness. Whereas stars (much like the Sun, but further away) produce their own light, the Moon and the planets (see Chapter 7) do not. We are only able to see those parts of the Moon that are illuminated by the Sun.

Light is all around us.

The idea that light simply hangs around rather than emanating from a source can persist into adulthood. The link between the Sun and the pattern of day and night is explored in more detail in Chapter 7. The need for a light source if there is to be light can be demonstrated in a blacked-out room with a torch. Challenge the children's understanding with questions to make them consider their ideas in greater depth: *If light is all around us, why isn't it under the bedclothes at night? Why do you need to turn the light on in the cellar?* Sometimes it is difficult to make the link between light and its light source: the Sun may be behind clouds, or it may be light in a room when the Sun is not shining directly through the window. In these cases, light has been scattered into our eyes from objects above or around us.

Questions

Where does the light come from on a cloudy day when you can't see the Sun?

If the child has already grasped that light comes from a source and travels in a straight line, he or she may well wonder how light can reach you when you can't see its source directly. But even on a sunny day, if you are standing in the shade and cannot see the Sun, it is still light. Firstly, the light can reach you by bouncing off other things (scattering or reflecting – usually scattering). Secondly, when the light source is obscured by something translucent (such as pale clouds), the light can still get through even though the image of the source is hidden. Frosted glass and paper lampshades are examples of this: because the material is translucent, the light is scattered (or diffused) through.

Teaching ideas

Sources of light (exploring and sorting)

Provide a collection of artefacts: some light sources and some non-luminous objects with a link to light (such as mirrors, lenses, bicycle reflectors and reflective strips). Children can examine and discuss these, then sort them into 'light sources' and 'not light sources'. Some objects can be tested by placing them in a lightproof box which has a peephole cut into it.

NB Naked lights such as lit matches or candles, if used at all, must NOT be placed in a box. The colours of naked flames are very interesting to observe – but any such observation must be carefully supervised.

Light travels (exploring and explaining)

Children can explore a darkened room with torches. Try to get across the idea of a path *from* the light source *to* an object *to* the eye. Encourage them to explain where the light is coming from, and where it is going. They can record their understanding of how they see an object by drawing a picture with lines and arrows to show the path of the light.

Concept 2: 'Then light hits something'

Subject facts

Light and materials – transparent, translucent and opaque

Light can travel through empty space (see above); but when it comes into contact with objects (matter), different things may happen, depending on the material they are made from and the structure of this material. A **transparent** material (such as air or thin glass) allows light to go straight through with little disturbance. A **translucent** material (such as muslin or frosted glass) also allows light through, but not straight through. The light ray will be deflected and diffused (broken up) by the structure of the material, so that although light does pass from one side to the other, it is so 'messed around' that you are unable to make out any distinct image. An **opaque** material (such as wood or metal) allows no light to pass through it – which means that you can see the object, as opposed to seeing through it.

The distinctions between transparent, translucent and opaque materials are relative. There are degrees of transparency: the thicker a piece of glass, generally, the less transparent it is; your hand is normally opaque, but becomes translucent if a very strong light is shone through it. These properties of materials, along with some surface properties (strange things can happen as light goes into or comes out of materials), can lead to some interesting optical effects.

Shadows

A shadow is formed when an opaque object is placed between a light source and a surface: the shadow falling on the surface has the shape of the object. The area of the shadow is unlikely to be completely devoid of light, as in most practical situations light will be scattered off other objects into the shadow area. If you look around now, you will probably see many different shadows with various depths of darkness. So, although an opaque material will block the transmission of light, its shadow will only be in complete darkness if there is only one light source and there are no other objects or surfaces around to redirect light into the shadow – which is unlikely, if not impossible. So in reality, a shadow is formed where an opaque object stands in the way of a strong direct light.

Many windows have curtains or blinds that are open in the day (to let light in) and closed at night (to stop light getting out). The effectiveness of different fabrics in blocking light varies tremendously. Some are very opaque, making it difficult to tell whether it is night or day outside; others are merely translucent – able to stop voyeurs from seeing inside, but not able to stop the sunrise from waking you up.

Seeing things – light scattering and absorption

What happens to the light that is blocked by an opaque material? Where does it go? With most materials, one of two things happens: the light can be **scattered** (it bounces back off), or it can be **absorbed** (it is converted to heat within the material). Usually what happens it is a combination of the two: brighter objects scatter more of the light and duller, darker ones absorb more (see Figure 2).

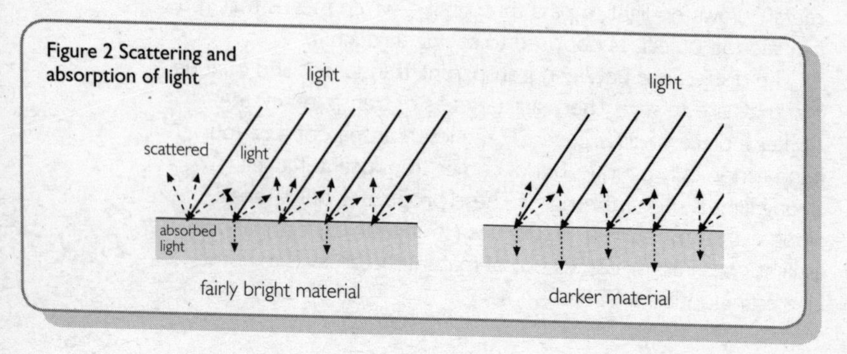

Figure 2 Scattering and absorption of light

It is important not to confuse scattering and **reflection**. Scattered light is returned from a surface in a jumbled state, so that the surface gives no image of the light source. Reflected light is returned in precisely the same quantities and positions to give an image of the light source (see page 111). For example, light is scattered from a white, plastic teaspoon, but it is reflected from a polished metal one (you can see yourself in it).

Light that is absorbed into materials becomes heat: when something is left out in bright sunlight, it tends to heat up. The duller or darker the object is, the more it heats up, since more of the light that falls onto it is retained as heat. A dull black object will absorb most of the light that hits it; a shiny white object will scatter most of the light that hits it, and will thus be easy to see.

If you have a reasonable grasp of things so far but don't want to push it, you can skip to the next paragraph. Still with me? Good. When the photons or particles of light hit a material, they can be either re-emitted (scattered) or absorbed. These photons are energy bearing. When they are absorbed, their energy is transformed from light to heat and thus heats the material.

In summary, the following chain of events leads to our seeing an object. Light is emitted from a source. The light strikes an object. The light is scattered from the object. Some of this light is scattered into our eyes. We see the object.

The different colours which most of us see depend on the proportions of the light scattered and absorbed by different objects. Colour vision is explored further under Concepts 3 (see page 120) and 4 (see page 126).

Reflection

As suggested above, reflection is a special way in which light can bounce off the surface of a material. Instead of being scattered, the light rays return at angles equal to the angles at which they hit the material – rather like a snooker ball bouncing off a side cushion (see Figure 3). Not only light coming directly from a source, but light scattered from an object, can be reflected in this way. If the reflective surface is smooth enough, a clear image of an object can be seen (see Figure 4). Mirrors are highly reflective surfaces that return exact images.

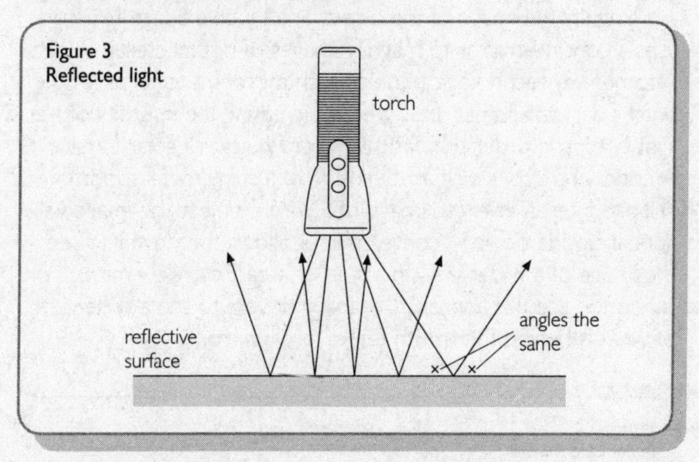

Figure 3
Reflected light

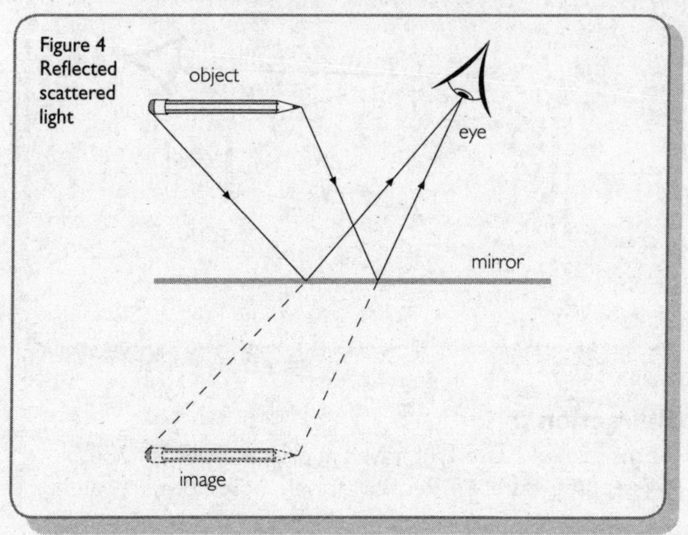

Figure 4
Reflected scattered light

Curved mirrors

A flat mirror can give an exact reflection of an object. If the surface of a mirror is curved, the image produced can be distorted: it may be shrunk, stretched, or inverted, depending on the curvature. A concave mirror (see Figure 5) can produce a 'stretched' image. A convex mirror can produce a 'squashed' image (see Figure 6). If you compare the path that the light takes in both diagrams, you will see that the angle at which it reflects from the mirror is the key factor in determining the size of the image.

The greater the curvature of the concave mirror, the more it will magnify the image of the object. The Hubble Space Telescope uses a concave mirror to magnify images of distant stellar objects. Another key technological use of such mirrors is to focus light onto a particular small area, greatly increasing the intensity of the light energy, in order to heat things very quickly: a 'solar furnace'. A concave mirror (such as the inside of a shiny metal spoon) will produce an inverted image of a *distant* object: the image will appear upside down. A convex mirror allows the viewer to see the image of a wider view in a smaller area. Rear view mirrors on cars often use this approach, enabling drivers to see a wide view of what is behind them with only a small mirror.

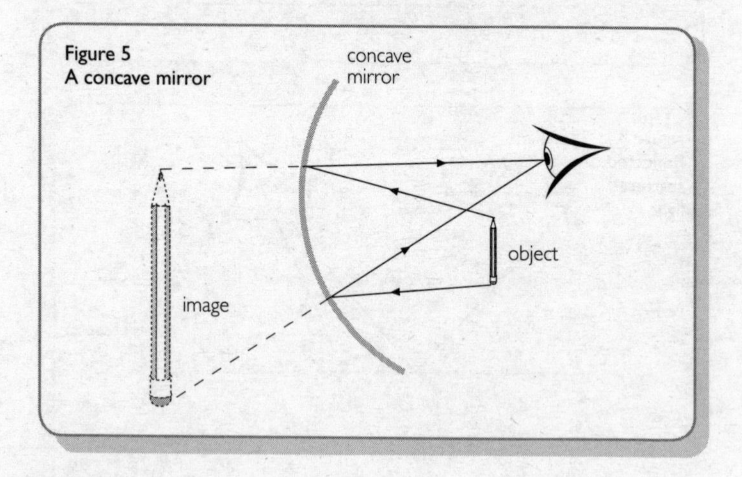

Figure 5
A concave mirror

Refraction

As mentioned above, light travels at different speeds through different transparent materials. It travels more slowly in water

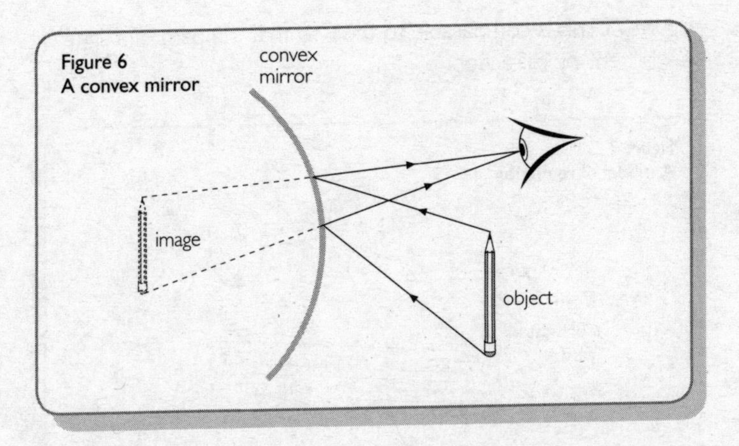

Figure 6
A convex mirror

than in air, and more slowly still in glass and perspex. In general, the denser the transparent material, the more slowly light travels through it. So if light is travelling through air and then enters some glass which is in its path, it will slow down; but once it has come out the other side, it will travel at the speed that it was going at before it entered the glass. Not many science textbooks, including this one, explain how that works! But knowing this property helps to explain why light sometimes appears to 'bend' – that is, how it can be **refracted** by glass or water, as in the 'bending pencil' effect (see page 114–5).

A model of refraction

The following model will give you a sense of what happens in refraction. Find a wide, smooth ramp and things to prop it up with. Next, join a couple of wooden wheels (of the same size) with an axle made from a piece of dowel. Let this simple vehicle roll down the ramp. Now place a strip of corrugated card across the ramp and let the vehicle roll down once more: it will slow down as it hits the strip of corrugated card, but will keep going in the same direction and then speed up again when it is past.

If you now place the strip at an angle across the ramp, the wheel that hits the strip first will be slowed down, making the vehicle veer off in that direction. The other wheel will then hit the strip and slow down. The first wheel to hit the strip will also be the first to leave and speed up, causing the vehicle to straighten up again. So it continues down the ramp, but further over to one side than it would have been had the strip not been there (see

Figure 7). This is comparable to the way that the path of a light ray is 'bent' by refraction.

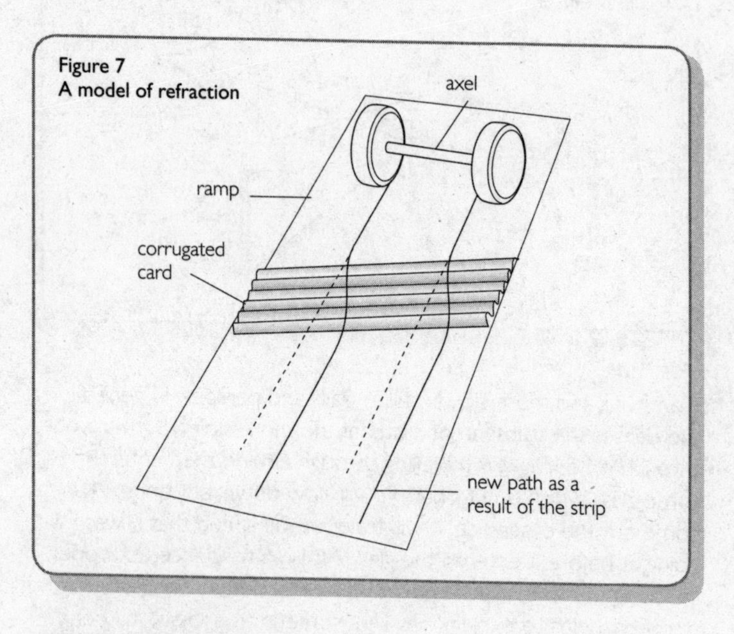

Figure 7
A model of refraction

Water refraction

If you slip a pencil end-on into a glass of water, it appears to bend (see Figure 8). This is an effect of refraction. The underwater end of the pencil appears to be nearer to you than it actually is, because the light from that end of the pencil is refracted by the water. More of the light from the pencil is directed towards your eye, so the pencil appears bigger. Because the light is going from water to air, the 'bent' pencil does not appear to straighten.

You can do some interesting refraction experiments using a clear plastic one-litre box and water. For example, with the box two-thirds full, place a pencil along the far edge going down into the water and view it from the diagonally opposite edge. You will see two pencils: one in each side. Now, using less water, tip the box over slightly and rest it on an edge with the pencil underneath. Twist the pencil around: you will see it curve!

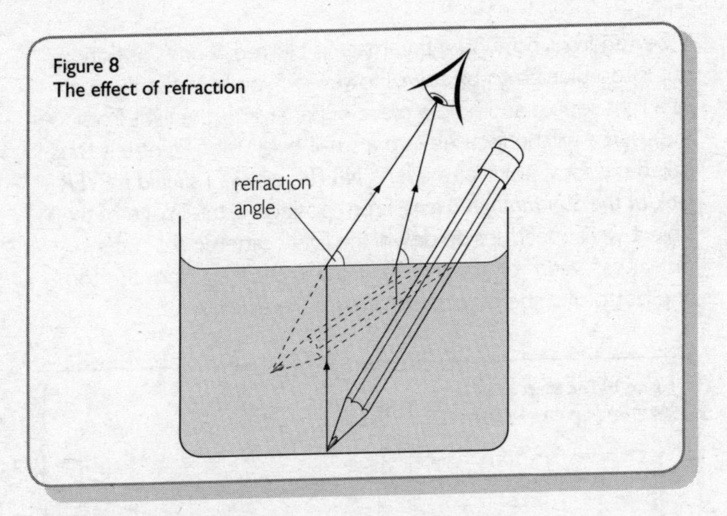

Figure 8
The effect of refraction

Refracting lenses

This is not something you would normally cover in primary work; but it would be an interesting topic for keen Year 6 children to explore. As concave and convex mirrors can change the size of an image, so can concave and convex pieces of glass (or any other transparent material). A clear plastic bottle, filled with water, acts as an effective lens and will do most things that a biconvex lens (one that is convex on both sides) will do.

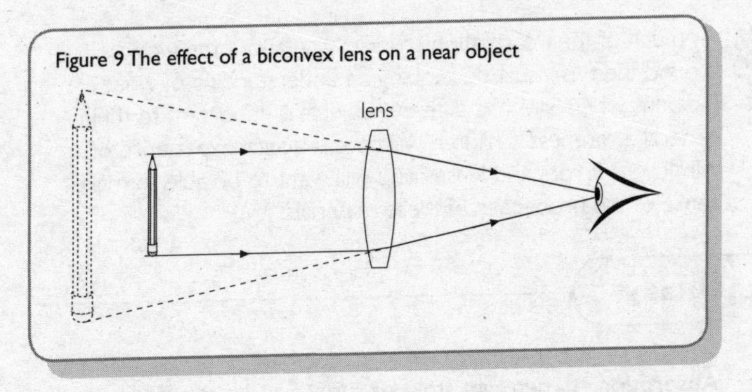

Figure 9 The effect of a biconvex lens on a near object

The light scattered from an object is refracted both when it goes into a biconvex lens and when it comes out. The image of an object that is close to the lens will be magnified (see Figure 9). For an object further away, what is seen will depend on how far the viewer is from the lens (see Figure 10 overleaf). If the object is viewed from position *a* the image is blurred. From position *b*, it is too blurred to be seen. However, if you hold the lens up to a light source and hold a piece of paper at *b*, the light from the source will be focused into a small, bright dot. Point *b* is thus called the focal point of the lens. ***NB This why you should NEVER look at the Sun through a lens.*** From position **c**, the image of the object will appear upside down: the light scattered from the object has been refracted so far that rays coming from the top and bottom of the object have crossed over.

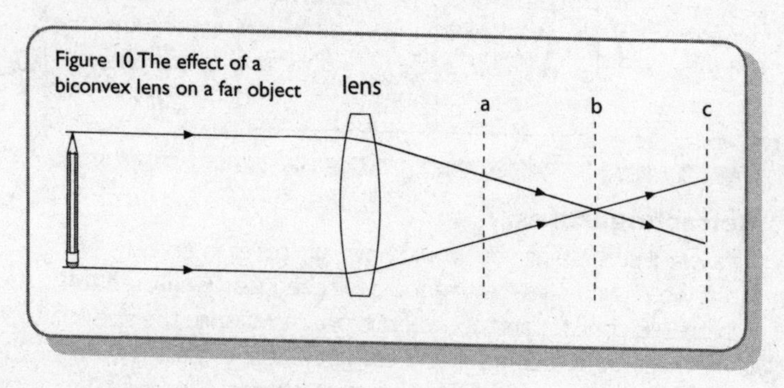

Figure 10 The effect of a biconvex lens on a far object

Why you need to know these facts

As much of the information children gain about the world around them is visual, developing an understanding of what happens when light hits different objects is important to their general awareness. Children will have practical experience of windows, mirrors and lenses, and will want to be able to make sense of the properties of these materials.

Vocabulary

Absorption – when light strikes a surface and is retained within it.
Opaque – a material which blocks the passage of light.
Reflection – when an image is returned from the surface of an object.
Refraction – the 'bending' of light when it passes from one transparent material to another.

Scattering – when light is returned from a surface.
Translucent – a material through which you can see light, but not an image.
Transparent – a material through which you can see an image.

Amazing facts

Most materials, even transparent ones, can produce some kind of shadow. On a sunny day, when the sunshine is streaming through the window onto the floor, open a window so that some of the sunshine is going directly onto the floor and some is going through the glass. Compare the brightness of the floor in the two areas. If glass were perfectly transparent, it would be invisible. Can you imagine a glass-shaped amount of wine apparently floating above the table?

Common misconceptions

You can get a reflection from any shiny object.

A reflection is really an image. Some 'shiny' surfaces scatter light very effectively, but do not reflect. Some surfaces can produce a reflection if you catch them at the right angle in the right light. 'Reflectors' on cycles and cars scatter light, but in a particular direction.

Shadow misconceptions.

Asking the children to draw a picture which includes the Sun, a tree and the tree's shadow should indicate how well they appreciate the relationship between a light source, an opaque object and a shadow. There are three crucial questions to ask as you look at the picture: *Is the shadow on the opposite side of the object from the light source? Is the shadow a similar shape to the object? Does the shadow touch the object?* If these questions are not all answered with 'yes', a walk outside on a sunny day will be helpful. I have often watched five-year-olds sprinting along the playground and stopping suddenly with the expectation that their shadow will keep going – as in a cartoon! Empirical evidence really does help to reinforce correct ideas.

Questions

Why am I upside down when I look into a spoon?

It has to do with the curved (concave) surface of the spoon and the way the light from your face is reflected in it. If you hold the spoon close to your face, the image of your chin appears near the top of the spoon and the image of your forehead near the bottom; so your face appears upside-down (see Figure 11).

Figure 11 Reflections in a spoon

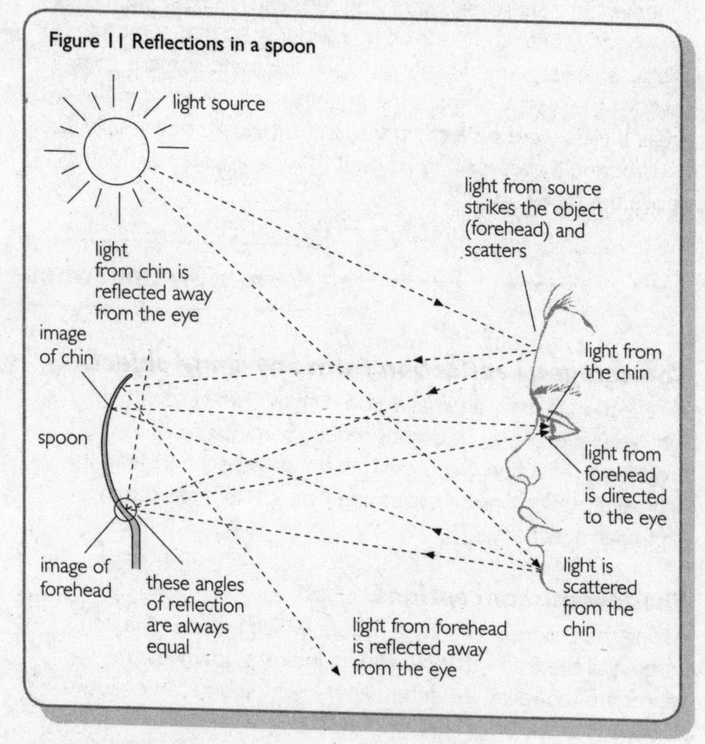

Why do rainbows appear in the sky?

The answer 'because sunlight is refracted through droplets of water' is only useful if you know about refraction! The best way to approach this question is to produce a spectrum in the classroom using a strong light source (perhaps the Sun), a tray of water and a mirror. Using a plant mister to spray water through a beam of sunlight is also a good way to produce a spectrum, and this is similar to what happens with rainbows: water droplets act as lenses, breaking white light up into a spectrum of colours.

The colours that appear in oily puddles and bubbles are due to the fact that the varying thickness of a layer of oil or bubble means that it reflects different colours (that is, wavelengths) of light.

On a hot day, why do objects in the distance appear to shimmer?

To understand this, you need to know about refraction. When light moves from a dense material to a less dense one (or vice versa), it is refracted and so changes direction. On a hot, sunny day, the air just above a sand, concrete or tarmac surface becomes heated. The heated air rises, because it is less dense than the air around it. So the light scattered from a distant object passes from denser air to less dense air. As it does so it refracts, causing the images of distant objects to shimmer. To explain this in terms that children will be able to grasp, it may be useful to compare it to the 'bent pencil'. If the water is still, you get a fairly stable image; but if you make a few little ripples, it appears to dance all over the place. The warm air rising creates a similar effect.

When I look out through a window, why do I sometimes see what's inside the room?

Transparent materials usually have smooth surfaces, which tends to make them reflective as well. The reflection from a glass window is more obvious if you view it from an acute angle, because you will not be blocking it. The reflection is also more obvious if there is more light on your side of the glass (being reflected) than on the outside (being let through).

Teaching ideas

Testing transparency (investigating and sorting)

The children can use a torch to test whether a sample of material is opaque or translucent. Can they tell when the torch is switched on by looking at it through the material? Point out to them that the thickness of the material and the brightness of the torch will affect the results. Of those materials that allow light through, can they say which are transparent (allow objects to be seen through them?)

Shiny and dull (sorting and ordering)

Ask a child to collect five to ten objects or materials from around the room and then place them in order of shininess. Record the order. Do the others agree with the order? Test the objects by placing them in a lightproof box (see page 108) and shining a light through a small gap: which one is the brightest now?

Bendy mirrors (exploring and observing at KS2)

The children can explore the effect on reflections of bending plastic mirrors, observing how the images change and trying to explain how the light is being reflected to make these images appear. A light box could be used to produce thin beams of light in order to trace the reflected paths.

Lenses and prisms (exploring and observing at KS2)

The children can investigate a selection of prisms and lenses in the same way as in 'Bendy mirrors'. Sectional convex or concave prisms are easier to use with light boxes than disc-shaped lenses.

Refracting prisms (exploring and explaining)

The children can take a regular-shaped lump of glass or perspex with flat faces (they usually come as triangular-based or square-based prisms), and look through it. If they rotate it, they will see images changing position, disappearing and reappearing. Can they tell you where the images are coming from? (More information about prisms is given below.)

Concept 3: Light and colours

Subject facts

The brightness of light

Light varies in intensity: light sources can emit smaller or greater amounts of light energy (or numbers of photons). The brightness of a light source can be reduced by reducing the energy input (for example, using a lower-voltage battery with a bulb), or by blocking some of the light using translucent materials such as tissue paper. Sometimes the brightness of one light source will mean that other, less bright light sources cannot be seen, such as the stars in daylight (see Chapter 7).

The colour of light

The cover of my diary is red, the ring binder is yellow, and my jeans are blue...but why? They are all illuminated by the same bulb, so why are they different colours? Light comes in many different colours, and scientists explain this fact in two different ways.

Using the 'wave' theory of light, which is the more familiar and popular approach, you can say that each colour is a slightly different wavelength of light: violet has the shortest wavelength (highest frequency) and red the longest (lowest frequency) within the spectrum of light that we can see. Outside this range are ultraviolet and infrared light. Using the 'photon' theory of light, you can say that colours are the result of photons having different energy levels, with red at the highest energy level and violet at the lowest. But it is enough to say that light comes in different colours!

White light

There is really no such thing as **white light**: there is no wavelength (or energy level) which is white. 'White light' is what we see when all the colours of light are mixed together in roughly equal quantities. Most of the light sources that we rely on, such as the Sun and domestic electrical lighting, provide us with white or near-white light. Street lamps tend to use cheaper, longer-lasting light sources that produce yellow 'sodium' light, making use of a more limited range of wavelengths.

Mixing colours

Because of the way that we see colours (see below), it doesn't require all of the colours of the spectrum for us to see an object as white. There are three **primary colours** of light that we can use to produce white. By varying the relative intensities of these three colours, we can make all the other colours. The primary colours are red, green and blue. By mixing red and green (that is, by shining these coloured lights on a white surface), we can make yellow; mixing green and blue gives cyan; mixing red and blue produces magenta.

Coloured objects

Most objects scatter the light that hits them, so why does something yellow appear yellow? As was explained previously (see page 109), when light hits an object it will be either absorbed or scattered. The relative proportions of light absorbed and scattered varies for different colours of light. When white light falls onto a banana, blue light is absorbed and red and green light is scattered (remember red + green = yellow). Grass absorbs red and blue light from white light falling on it, and scatters green light.

The key point is that we see different colours according to which colours of light different pigments scatter or absorb. A white object scatters all the colours of light. So does something that is grey, but it also absorbs more of all the colours as well. Any colour appears in a darker shade if more of the light is absorbed.

Rose-tinted spectacles

Rose-tinted spectacles (or ones with any other colour of lens) affect how we see the world. Red glass or (transparent) plastic absorbs all colours of light other than red. Some red light is scattered from it (which is why it looks red), but it will only allow red light through it. Objects scattering white light will thus appear to be red, and so will objects scattering red light. Objects that scatter some red (such as magenta or yellow objects) will appear red, since blue or green light will not get through the glasses. Because green and blue objects don't scatter any red light, they will appear to be black.

Why you need to know these facts

Children can normally see a range of colours, shades and degrees of brightness. They need to be able to explain how this is possible.

Vocabulary

Primary colours (red, blue and green) – the colours of light that our eyes are able to detect.
White light – a mixture of all the colours of light (though for our eyesight, we only need to mix red, blue and green to make white).

Common misconceptions

The most common colour of light is white.
Red, green or blue light is light at a particular wavelength. White light cannot be described in the same way. It is all the colours of visible light combined in (more or less) equal intensities. What we see as white light is thus a particular mixture of different colours.

The primary colours of light and paint (pigment) are the same.
This assumption can be tested if you have a room dark enough to mix lights. If possible, it is best to use theatre lamps with red, blue and green colour filters. The control panel should allow you to dim each light independently, mixing the three colours to produce other colours of light. The primary colours of pigment are yellow, cyan and magenta (the secondary colours of light). These colours, along with black, are the ones used in most colour printing processes (for example, the inks in a computer printer). When mixed, cyan and yellow will produce green; magenta and cyan, blue; and yellow and magenta, red. However, all three together will only produce grey. If you try mixing the same colours in paint, the colours produced will not be the same. Light adds colours; pigment subtracts colours.

Teaching ideas

Splitting white light (demonstrating, exploring)
White light can be split into a spectrum of colours by means

of refraction (see page 121). To explore this, the children will need a strong white light source (such as the Sun), a prism and a screen to project the spectrum onto. Figure 12 shows how three possible types of 'prism' can be used to produce a spectrum: a commercial perspex 60° prism; a mirror in a water tray; or a clear one-litre plastic box partly filled with water. The spectrum produced by these methods will be a band of colour, gradually changing from blue-violet at one edge through green to red at the other. Although different colours can be seen, there are no distinct colour 'bands' within the spectrum.

Figure 12 Splitting white light

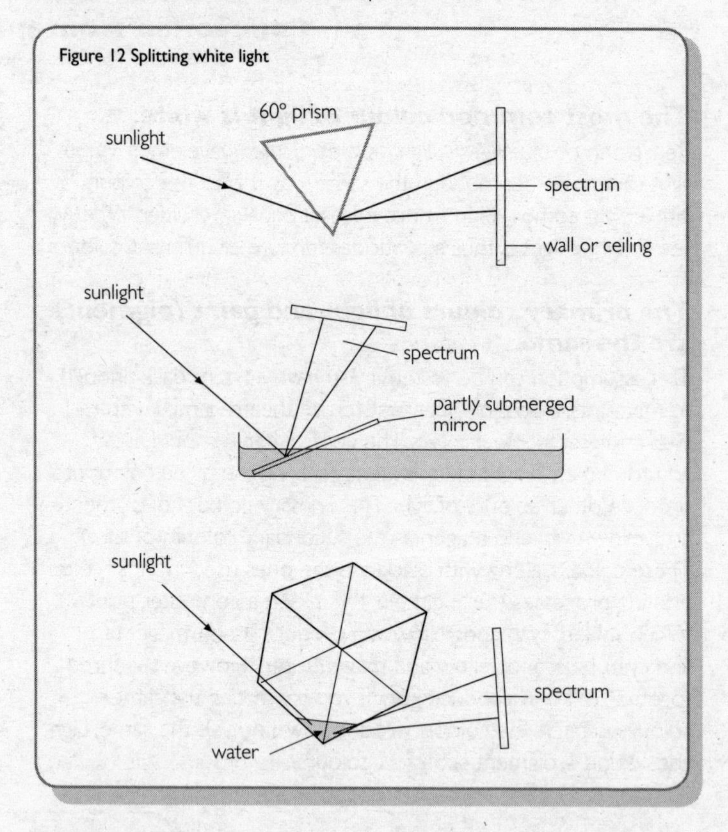

Ask the children to describe the gradual colour changes that they can see and try to identify the order. Why can't they see magenta? (Because red and blue are at opposite ends of the spectrum and so do not mix.)

Safety at night (testing and sorting)

Provide a collection of 'safety clothing' with highly fluorescent or bright surfaces. The children can test which of them are most visible in the light box (see above). They could make an orange light (similar to the colour of street lamps), then test different materials first with white light and then with orange light. Ask them: *Which colours scatter light best? Which colours absorb light best?*

Colour mixing (observing, using ICT)

With adult supervision, the children could investigate the effects of shining theatre lamps with blue and green filters on a white screen in a darkened room (see page 122), or use art software on a computer to experiment with mixing primary colours of light at different intensities.

Secret messages (observing)

This fun activity helps children to remember the primary colours of light, and demonstrates the effect of viewing different colours through sheets of transparent coloured plastic (known as **gels** or **colour filters**). A red filter works best, but others can be used to show the different effects. The children can use the figure [8], copied from a calculator display, as a template to write block capital letters in a message. They should use blue and green pens for lines to be seen and red, yellow, orange and pink pens for lines to disappear when the message is viewed through a red filter. This is a great activity for an 'indoor break', and can result in excellent displays.

Rose-coloured spectacles (observing)

The children can find out how the world looks when viewed through coloured gels. They could try using a different colour for each eye. Red and green work quite well together. Seen through the red gel, white and red objects appear to be the same colour, and blue and green both appear black. Seen through a green gel, white appears to be green and red appears to be black. The two eyes will see things very differently from each other!

Concept 4: How we see

Subject facts

The human eye

How the eye works is more a biological than a physical topic; but since the eye's functioning depends on most of the principles outlined above, it is worth discussing briefly here (see Figure 13).

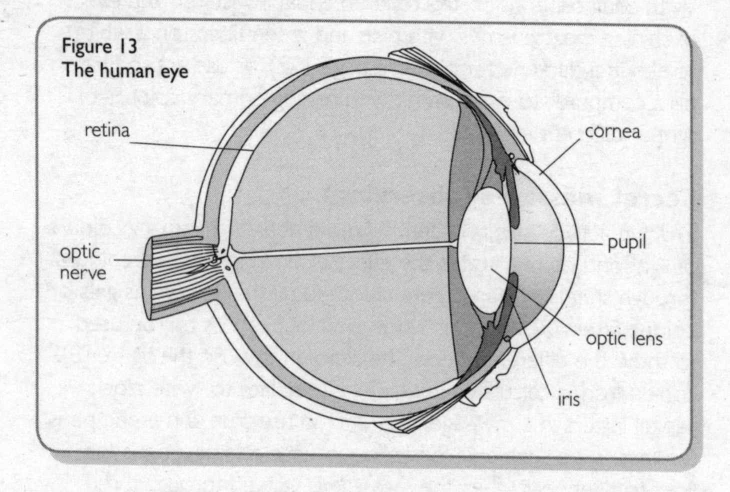

Figure 13
The human eye

A tough surface, the **cornea**, covers the front of the eye. Light enters the eye through a 'window', the **pupil** (which is the black central part). The coloured ring around the pupil is the **iris**, which can become bigger or smaller to adjust the amount of light allowed into the eye. In darkness or dim light the iris becomes thinner, allowing the pupil to open wider. In bright light the iris becomes wider, almost covering the pupil. Behind the pupil is a biconvex, transparent liquid-filled sack: the **optic lens**. Tiny muscles stretch or squash the lens, making it thicker or thinner so that images of different objects can be focused onto the back of the eye. The lens becomes thicker when focusing on near objects.

The inside back surface of the eye, the **retina**, is covered with light-sensitive cells. These cells respond to red, blue or green light (which is why we only need those three colours to make white), and transmit signals to the brain via the **optic nerve**. It is the brain that actually 'sees': it makes sense of the images, and interprets the relative intensities of red, blue and green light as the various colours that we see.

Visual overload

When the light is too bright, the iris expands to restrict the amount of light getting into the eye. This can be inconvenient if you are passing directly from bright illumination to a dimly lit place – for example, if you go from a sunlit garden into a house, you will be 'in the dark' for a few seconds until your irises open up to let more light in. In reverse, when you go from a shaded place to a bright place, you are temporarily dazzled until your irises reduce the amount of light entering your pupils (in order to protect the retina from overexposure).

Your eyes can also adjust to an excess of one colour. The brain works on the principle that there is a range of colours in your field of vision, and that these change constantly as your eyes 'wander about'. If one colour is much brighter than the others, or stays in a particular area of your vision for too long, your brain begins to say 'That can't be right, better turn it down.' If you stare at a green cross on a white background for thirty seconds, then blink away and quickly look at a plain, white background (blinking a couple more times if necessary), you will see an **after-image** of a magenta (pink) cross. This is because your brain has 'turned down' the green receptors to avoid overload while looking at the green cross: when you look at a white background, your brain 'tunes out' a cross-shaped patch of green from the white – and so you see a magenta cross (made up of blue and red).

For this reason, the furniture in operating theatres tends to be draped in blue or green cloth. After staring into a blood-filled cavity, you will tend to go 'red-blind' and be unable to make things out. Looking away at something green or blue will bring your eyes back into colour balance.

Why you need to know these facts

Eyes are our means of detecting light, and an understanding of how they work will increase children's ability to appreciate the science of life.

Vocabulary

After-image – negative image 'seen' as a result of parts of the retina closing down after overexposure to light or particular colours of light.
Cornea – the tough outer covering of the eye.
Iris – opens and closes to let more or less light into the eye.
Optic lens – the clear opening that allows light into the eye.
Optic nerve – transmits the information received by the retina to the brain.
Pupil – the black centre of the eye that allows light through to the lens.
Retina – the internal surface of the eye that senses light.

Common misconceptions

As children, we soon become aware that we use our eyes for seeing things, long before we have any scientific understanding of vision or the eye. How the eye works and how we use light to see things are frequently misunderstood. The early Greek philosophers had two conflicting theories about how we see things. The followers of Pythagoras believed that the eye contains light and the light comes out when we look at things. The alternative (and ultimately correct) view was held by the followers of Democritus: that light travels from a source to an object and then into the eye.

You see things because light comes out of your eyes.
This classic misconception can be overcome by asking the child to go into a darkened room and find something. 'I can't, it's too dark in here!' Light is clearly not coming from the child's eyes!

When light is in your eyes, you can look around and see things with it.
Repeat the 'darkened room' test, and point a small torch at the child's face. The child will see the torch, but not the object. Ask where the torch needs to be pointed. 'At the thing I want to find'. By this point, the child should recognise that the light goes from the source to the object to his or her eyes.

Questions

Why do I see spots when I blink after looking into a torch light?

Your eyes try to protect the light-sensitive cells of the retina from damage by blocking out very bright lights. First the black bit (the pupil) in the centre of your eye gets smaller to let less light in. Then the cells in the back part of your eye (the retina) start to close down, because they are getting overloaded. When you look away, the 'shut down' patches of your retina cause temporary dark spots on your vision. ***(NB Warn children not to stare at very bright lights, because of the damage this will cause to the retina. They should be especially careful when emerging from a dark room into bright light.)***

Why do people sometimes need glasses?

When you look at things close to you and then further away, the lenses in your eyes have to refocus by changing shape slightly. If you focus on your hand in front of your face, everything else becomes blurred; then if you focus on something further away, your hand becomes blurred. If you can't make the lenses in your eyes change shape enough, the lenses in glasses can help you. Glasses have to be specially made for each person's eyes. ***(NB Warn children not to try on each other's glasses: this will cause blurred vision, as well as headaches.)***

Teaching ideas

Opposite pictures

This activity can be linked to art work. Ask the children to use bright paint or collage materials to produce pictures that are only seen in the correct colours when you stare at them for a while and then look at a sheet of white paper to see the afterimage.

To find the 'opposites' of the colours they need, the children can make a colour wheel (red, blue and green with the secondary colours of light painted between them) to use as a visual reminder (see Figure 14 overleaf). This could be made using a drawing or painting software program.

Alternatively, the children can stare at a 'true' colour version and then see its 'opposite' after-image on a white background.

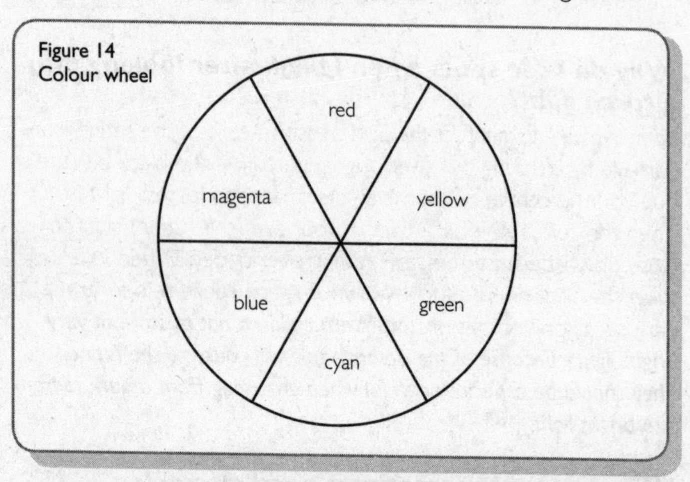

Figure 14
Colour wheel

Resources

The following will be useful for practical work on light:
- various coloured plastic drinks bottles;
- a selection of prisms;
- a selection of lenses (concave and convex);
- a selection of mirrors (plastic glass; flat, convex and concave);
- a selection of fabrics and other materials (opaque, translucent and transparent);
- electric torches;
- coloured theatre gels (lamps);
- colour paddles or tetrahedra.

Websites

www.animations.physics.unsw.edu.au/jw/light/color-mixing.htm#3
www.bbc.co.uk/schools/scienceclips/ages/10_11/see_things.shtml
www.optics4kids.org/Home.aspx

Books and CD-ROMs

Scholastic Primary Science: Sound and Light
Investigate: Light (Scholastic) – exciting non-fiction readers

Sound

Chapter 6

Key concepts

Three key properties of sound are of particular interest to primary scientists:

1. Sound travels from a vibrating source through a medium.
2. Sound can vary in loudness, pitch and timbre.
3. Sound can be detected by our ears.

Sound concept chain

See 'Electricity concept chain' (page 10) for general comments.

KS1

There are many different sounds around us. Every sound has a source. As you move further from the source of a sound, it gets quieter. Sounds can be made in different ways (for example, by hitting, shaking, scraping, blowing or plucking a musical instrument). We hear sounds using our ears.

KS2

Sound travels through the air. Vibrating objects are the sources of sounds. A sound can be made louder by making the object vibrate more. Some objects can be made to vibrate faster or slower in order to make the note they produce higher or lower. There are various ways of changing the rate at which an object vibrates. Sound vibrations can travel through many different materials. Sound can travel through some materials better than others. Sound may be reflected (echo) from hard, solid objects.

KS3

Sound cannot travel through empty space (a vacuum). The greater the amplitude of a vibration, the louder the sound it causes. The higher the frequency of a vibration, the higher the pitch of the note it produces. The ear has different parts which combine to enable us to detect sounds. Each person has a unique range and quality of hearing. Hearing can be damaged (either temporarily or permanently) by loud noises. Technology is available to repair or improve some forms of hearing deficiencies. Sound penetrates some materials and is reflected from others. Reflected sound can be used to detect the position and shape of a solid object (sonar, ultrasound).

Concept 1: Sound transmission

Subject facts

There are a tremendous variety of sounds around us at all times. They differ in a number of subtle ways, but they are all produced by something vibrating. In many cases, the exact cause of the **vibrations** is quite difficult to identify. However, there are many sound-producing objects where the source of the vibrations is open to easy examination – for example, musical instruments which are plucked or struck (guitars and drums). The more interesting problem is explaining how these sounds travel out from the source.

Sound waves

The term 'sound waves' can give the wrong mental picture. They are not like ripples on a pond, undulating up and down (**transverse** waves) and radiating out from the point where a stone fell. Imagine what a vibrating drum skin would look like if it were shown in slow motion: it would be moving in and out (see Figure 1). As it moves out it pushes against the surrounding air, compressing it and making it 'bump' into the air next to it. A useful analogy is a car running into the back of a row of stationary cars, causing them to knock into each other.

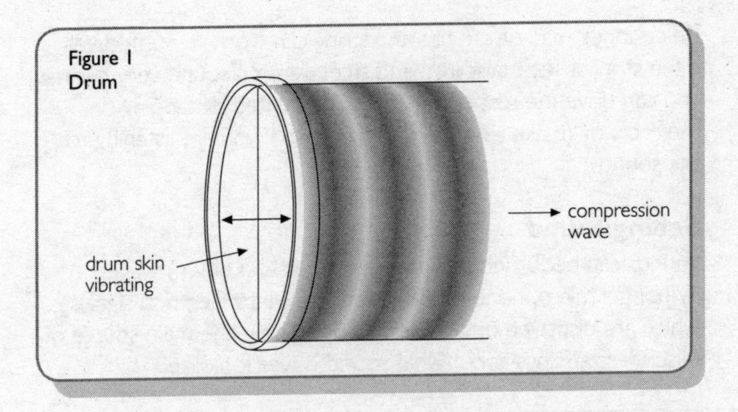

Another way of modelling sound is to hold one end of a Slinky® and repeatedly flick the other end (see Figure 2): this will send compression pulses along its length (longitudinal waves). The pulse might even bounce back from the end, as in an **echo**.

On reaching the ear, these sound pulses press on the eardrum (see Concept 3, page 140), causing you to hear them.

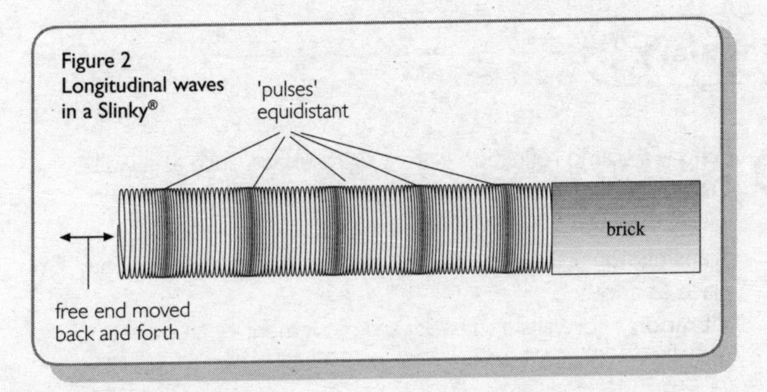

How sound travels

Sound pulses or vibrations can travel through most materials. Some materials are better conductors of sound than others, because the molecules are arranged in a way that makes vibration easier. Sound and light are very different in this respect: sound needs a medium to travel through, but light doesn't (see page 105). Whales can communicate with each other over hundreds of miles by sound pulses travelling through the sea.

'Telephones' made from tins and string can work very effectively if the string is kept taut. Iron and steel transmit sound very clearly: you can drive the rest of the school demented by tapping on the radiator (make sure it's not turned on if you are listening to the sounds).

Seeing sound

Sounds can be turned into transverse (up and down) waves by using a microphone to transform them into electrical signals which are displayed on an **oscilloscope**. This is the main source of misunderstandings about what sound waves 'look like'.

Why you need to know these facts

An appreciation of how sound is transmitted is fundamental to understanding the nature of sound. All further work in this area will be supported by a secure introduction here.

Vocabulary

Echo – a sound reflection from a rigid surface, such as a wall.
Oscilloscope – a device for displaying in a visible form sound waves that have been converted into electrical pulses.
Transmission – the process by which sound travels from one place to another.
Vibration – forward and backward movement of an object (usually very rapid).

Amazing facts

In the air at sea level, sound travels at 340 metres per second (ms^{-1}). It travels faster in water ($1400ms^{-1}$) and faster still in steel ($5000ms^{-1}$).

Common misconceptions

Sound transmission

There are relatively few common misconceptions about sound, though misunderstandings can arise from the use of vocabulary. For example, the question *How do you hear that sound?* might be answered 'Because I listen carefully!' A better question to ask would be: *How does the sound get from the drum to your ear?* The idea that jagged lines of sound come from a noisy object (as in cartoons and comics) may confuse some children.

Sound production

Sometimes children find the link between the vibration of an object and sound difficult to make. Holding an inflated balloon in front of your face and singing, humming or speaking at it will cause it to vibrate, demonstrating that sound and vibration are essentially the same thing.

Questions

Why does the sound of a car change as it goes past?

As a speeding car comes towards you, it is 'catching up' with its own sound, so the sound pulses you hear are squashed closer together (higher note). As it goes away from you, the sound is 'left behind', so the pulses are drawn further apart (lower note). (See Figure 3 overleaf.) As a model, imagine a conveyor belt running towards you with a person standing beside the belt further up placing chocolates on it at regular time intervals. If this person were to walk towards you while still placing the chocolates, you would start to receive chocolates more often. If the person were to walk in the opposite direction, the chocolates would arrive less often.

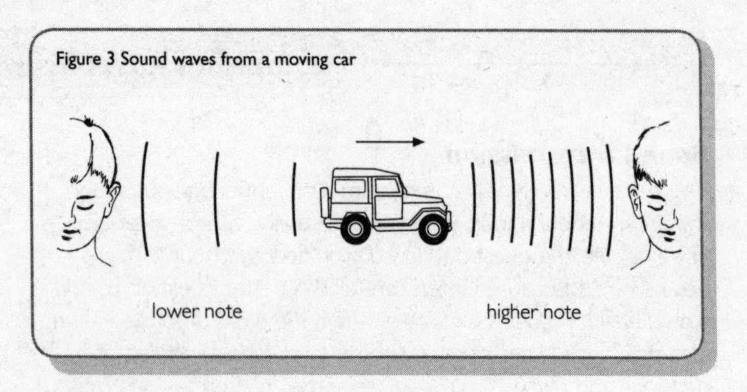

Figure 3 Sound waves from a moving car

Teaching ideas

What happens when an aeroplane travels at the speed of sound?

The sound pulses build up in front of the moving aeroplane (a 'pile' of sound pulses all travelling together). As it goes past, you will hear all of these pulses at the same time: a 'sonic boom'. In the conveyor belt model, the person is walking towards you at the same speed as the belt, piling up chocolates which all reach you just as the person passes by.

What makes a good sounding board? (exploring)

Give each group of children a tuning fork. Ask them to take turns hitting one of the prongs against the tabletop (to start it vibrating), then place the base of the fork against different surfaces in the classroom (tabletop, wall, carpet and so on). *Which surfaces amplify the sound? What do these surfaces have in common?* (They are all hard.) Ask the children to record materials in two sets according to whether or not they amplify the sound.

Ask the children to 'listen' to a vibrating tuning fork by placing it against their head. Can they 'hear' the sound better in some places than in others? Invite them to make a 'sound map' of the head. **NB** If children can 'hear' the fork better when its base is placed against the bridge of the nose or the skull behind the ear than when it is held next to the ear, they may have a hearing problem and should be referred to a doctor.

Concept 2: Different sounds

Subject facts

Volume

Loudness, volume and **amplitude** are different terms for the same basic concept. Amplitude is a measure of the vertical distance between the 'peak' and the 'trough' of a sound wave as represented on an oscilloscope. The louder the sound, the more powerful the sound pulse: the sound wave (or pressure wave) of an explosion can sometimes cause as much damage as the flying debris. In general terms, the harder you hit a drum, the greater the vibrations are and the louder the sound.

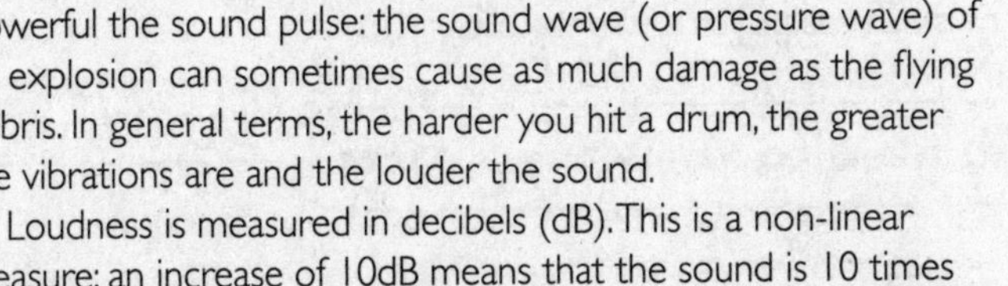

Loudness is measured in decibels (dB). This is a non-linear measure: an increase of 10dB means that the sound is 10 times as loud. A sound level of 0dB is near to absolute silence (30dB is a quiet whisper). 140dB is the level at which permanent hearing damage can be caused. The pressure of the sound waves at 140dB is 10,000,000 times greater than at 0dB.

Pitch

Stronger vibrations are not to be confused with faster vibrations. The faster an object vibrates, the higher the pitch of the note it produces – that is, the more frequent the sound pressure pulses are. Musical notes correspond to particular **frequencies** of vibration. For example, middle C is 256 vibrations per second, or hertz (Hz). An octave higher is 512Hz; an octave lower is 128Hz. In general, other things being equal, a smaller, thinner or tighter vibrating object will have a higher pitch. Some instruments can be 'tuned' to a particular pitch or frequency (for example, by tightening a guitar string).

A glass bottle partly filled with water can be made to produce a note in two distinct ways: hitting and blowing. If you strike it with a beater, you will notice that the key factor in determining the pitch of the sound is the amount of water: more water makes a lower note. If you blow across the mouth of the bottle, the air space is the important factor: the bigger the gap between the water surface and the lip of the bottle, the lower the note.

Tone

Very few vibrating objects produce a pure, single-frequency sound; a tuning fork is one that does. Different instruments may make the same note, but we are able to distinguish between them because of the tone or **timbre** of the sound. The tone is an effect of the range of secondary sounds that are produced along with the main note: those for a trumpet are different from those for a violin. Oscilloscope traces of a trumpet and a violin will look quite different, even though the frequencies of the notes being played are the same. This uniqueness allows our ears to identify subtle differences, such as recognising voices or knowing which musician has played a wrong note.

Why you need to know these facts

The terminology used here is also often used in everyday speech. A knowledge of the concepts underlying the words will allow the children to use this vocabulary much more effectively.

Vocabulary

Amplitude – the strength or intensity of sound vibrations, commonly perceived as 'loudness' or 'volume'.
Frequency – the number of vibrations per second corresponding to a given sound (in musical terms, its 'pitch'), measured in hertz (Hz).
Timbre – the collection of secondary notes that accompany the main note to add richness to the sound.

Amazing facts

- The highest note recorded in a human voice is 4340Hz, and the lowest is 20.6Hz – a greater range than that of a grand piano (7.25 octaves).

- When the volcano Krakatoa exploded in 1883, the sound was heard 3000 miles away and four hours later was described as being like the 'roar of naval guns'.
- A humpback whale can make a noise of up to 190dB, much louder than a jet aircraft taking off.

Common misconceptions

Pitch and volume

Children often confuse quiet sounds and low-pitched sounds. The vocabulary needs to be exemplified carefully to ensure correct usage.

Questions

Can you smash a glass with your voice?

Personally, no – doing the washing-up is my preferred method. But it can be done. All things have a frequency at which they naturally vibrate. Running a damp fingertip around the lip of a wineglass will (with practice) make it vibrate and produce a note. If this note (the resonant frequency of the wineglass) were sung back, the glass would vibrate. If the note were loud enough, the glass would vibrate so much that it would break.

How can you tune a guitar? (investigating)

Provide some guitar strings (or wires or threads), a G-clamp, some weights for hanging and three old rulers. Set up the apparatus as shown in Figure 4. Ask the children, working in groups, to test the effect on the sound of changing different variables: length of string, thickness of string, tightness of string, type of string. Emphasise that only *one* variable should be changed by each group. Ask the groups to say what relationships they have found.

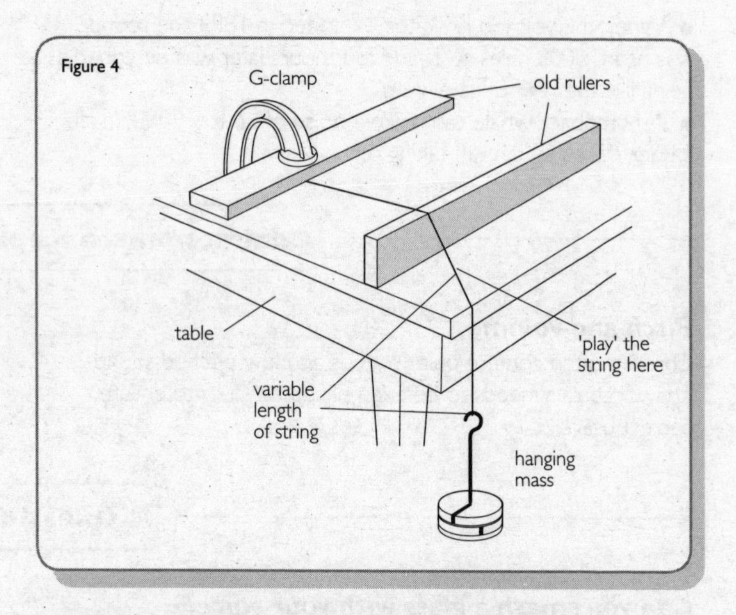

Blowing bottles (investigating and sorting)

Provide a large selection of bottles of various shapes, sizes and colours. What determines the note each bottle makes? Ask the children, working in groups, to explore the effect of changing different variables such as: size of neck, colour, capacity, height. They should order their bottles according to the variable – does this also give the order of lowest to highest note? (Height is the key variable which affects the pitch.) Ask them to draw the bottles in order of pitch, then predict where new bottles will fit in.

Concept 3: Sound reception

Subject facts

Although our ears are our main sound receptors, other parts of our bodies are also susceptible to sound vibrations. The sound of a dentist's drill is transmitted very effectively through the teeth. The bass at a rock concert is felt by the stomach as much as it is heard by the ears (in fact, several governments have developed 'riot control' weapons using loud, low-frequency sounds to stun or cause diarrhoea).

The human ear

Our ears provide for most of our listening needs. The human ear has three parts: a collector, a transmitter and a receiver (see Figure 5). The outer ear flap (**pinna**) collects sound pulses and directs them towards the **eardrum,** or tympanic membrane. Having ears that are tilted slightly forwards means that we hear things in front of us better than things behind us. The vibrations of the ear drum as sounds hit it are transmitted through the middle ear via three small bones (the hammer, anvil and stirrup) to the oval window at the entrance to the inner ear.

The **cochlea,** or inner ear is a cone shape rolled up like a snail's shell, with the widest end attached to the oval window. It is filled with fluid, and its inner surface is covered in small clumps of very short sensory hairs. Sounds entering the inner ear make the fluid move and bend the hairs, which transmit signals to the brain. Higher notes cause hairs further inside the cone to bend. The brain interprets these signals as 'sounds'.

Figure 5
The human ear

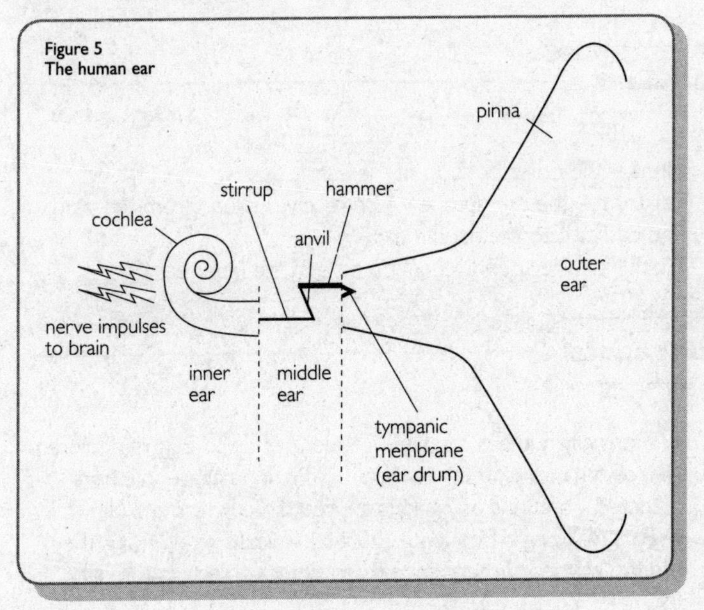

Hearing

Sounds can normally be heard in the range 20 to 20,000Hz, though the higher frequencies get harder to hear as you get older. The ear is most sensitive to sounds within the human vocal range.

Some hearing problems can be solved by technological solutions. More sound can be collected by using an ear trumpet to extend the pinna. A hearing aid can be used to increase the volume: the sound is picked up by a tiny microphone, amplified and then retransmitted through a tiny loudspeaker. If the middle ear is failing to transmit, it can be replaced by plastic inserts or bypassed by anchoring a hearing aid to the skull. In some cases, it is possible to send sound directly as electrical impulses to the cochlear nerves.

Why you need to know these facts

Ears are our sound receptors, so a basic understanding of how they work is important in linking the physical occurrence of sound to how we actually perceive it.

Vocabulary

Pinna – the outer portion of the ear.
Eardrum – the membrane which receives sound from the pinna and passes it to the middle ear.
Cochlea – the sound reception part of the inner ear.

Amazing facts

- Normally humans can hear frequencies up to 20kHz. Children with asthma can sometimes hear 30kHz. In tests, people have detected vibrations of 200kHz applied directly to the skull.
- Bats make high-frequency (90kHz) squeaks when flying: they use the echoes of these squeaks to detect objects in the dark.

Questions

Why can't I hear very well when I have a cold?

The mucus that blocks your nose can also block your ears! There is a direct connection between the two, which is why you feel pressure in your ears when you hold your nose and try to breathe out. The mucus allows pressure to build up, so the ear drum doesn't work as well and the sound vibrations are muffled as they travel through the middle ear.

Teaching ideas

Sound direction (exploring, modelling)

This activity is best done in a hall or gym. Blindfold a child and place him or her inside a circle of children. Point to a child in the circle, who clicks his or her fingers or tongue. Can the blindfolded child point to the source of the noise? Use this to discuss the importance of having two separate ears for locating the source of a sound.

Sound book (using ICT)

The children can use a presentation software package to present a collection of digitally captured sounds and images. They could make an electronic 'book' where the 'reader' clicks on images to hear the sounds that they make ('The cat goes meow'), or clicks on icons to hear demonstrations. For example, xylophone keys and recorded notes could be used to demonstrate the idea of pitch.

Resources

- Not surprisingly, most of the resources required for this topic are ones which make sounds. Children can explore how different toys produce noises. They can investigate musical instruments: a drum, a xylophone or chime bars; maracas; a stringed instrument (to be plucked and/or rubbed); and a recorder or swanny whistle. Lengths of plastic and/or metal pipe can be cut to different lengths (carefully sand or file the cut ends) and 'tuned'. Different-sized containers can be filled with different-sized particles (from sand to marbles) to make shakers.
- Tuning forks are useful – middle C (256Hz) in particular, as most musical instruments tend to be tuned to that note and the processes of 'halving' and 'doubling' (moving down and up by one octave) are relatively straightforward.
- Balloons vibrate well when sound waves hit them. A collection of bottles can be compared by blowing and striking.
- The children can use a stethoscope (or length of PVC tubing and funnel) to listen to each other's heartbeats.
- Several computer programs designed to teach a foreign language have voice comparison software. Most remote data logging/capture systems can be used with a microphone to capture sound patterns and record them on a computer.

Books and CD-ROMs

Scholastic Primary Science: Sound and Light
Investigate: Sound (Scholastic) – exciting non-fiction readers

Earth in space

Chapter 7

Key concepts

Ideas concerning 'Earth and beyond' are usually first taught during Key Stage 2; but they are fundamental to children's understanding of how their environment changes with time. Their development of this understanding is constrained by the scale of the physical entities involved. The key ideas to be developed are:
1. Many different types of objects make up the Solar System and the systems beyond.
2. The movements of the Earth, Sun and Moon cause periodic changes on the Earth and in the sky.

Earth in space concept chain

See 'Electricity concept chain' (page 10) for general comments.

KS1

The Sun is a source of light. The Moon is not a source of light, but is illuminated by light from the Sun. Daylight is caused by the Sun's light illuminating the Earth.

KS2

The Earth is a planet, and is approximately spherical. The Sun is the nearest star, and is a sphere. The Earth is in orbit around the Sun. There are seven other planets in orbit around the Sun, making up the Solar System (Pluto is now classed as a 'dwarf planet'). The Moon is in orbit around the Earth. The Earth rotates on its axis once every day. As the Earth rotates, a point on its surface alternates between facing towards and facing away from the Sun, causing day and night to occur. As the Earth rotates, the position of the Sun in the sky changes, causing shadows to change in length and direction. The Earth orbits the Sun once each year. Planets closer to the Sun than the Earth take less time to complete an orbit; planets further out take longer. Seasonal changes, including day length, are caused by the movement of the Earth relative to the Sun. The Moon takes approximately 27.3 days to complete an orbit of the Earth. The phases of the Moon (which repeat every 29.5 days) occur because only those parts of the Moon illuminated by the Sun are visible from the Earth.

KS3

Seasonal changes are caused by the inclination of the Earth's axis relative to the plane of its orbit around the Sun. The visible portion of the Moon is determined by the relative positions of the Earth, Moon and Sun. Stars in the sky cannot be seen in daylight due to the brightness of the Sun. Stars are similar to the Sun, only much further away. Many small planets, called asteroids, orbit the Sun. Seven of the planets in the Solar System have their own satellites. The Sun only rises exactly in the east and sets exactly in the west on the two days of the year which have equal lengths of day and night. The number of daylight hours varies throughout the year, being longest in midsummer and shortest in midwinter. Planets and satellites remain in orbit due to the effects of gravitational attraction. The combined gravitational attraction of the Sun and the Moon causes ocean tides on the Earth. The Earth moving directly between the Sun and the Moon causes a lunar eclipse (the shadow of the Earth falls across the Moon). The Moon moving directly between the Sun and the Earth causes a solar eclipse (the shadow of the Moon falls across the Earth).

Concept 1: Our place in the universe

Subject facts

Space

As Douglas Adams' *The Hitchhiker's Guide to the Galaxy* tells us, '*Space is big. Really big. You just won't believe how vastly hugely mindbogglingly big it is. I mean you may think it's a long way down the road to the chemist, but that's just peanuts to space.'*

Space is so big that we measure the distances in terms of how far light will travel in a **year** – and it travels at a speed of 299,792,458 metres per second. Light will take 8 minutes and 15 seconds to get from the Sun to the Earth. It will take about 5 hours and 25 minutes to reach the **dwarf planet** Pluto – which tells you that Pluto is a lot further away. But the Sun's light will take 4.2 *years* to travel to the next nearest star, which is thus said to be 4.2 'light years' away. Objects *millions* of light years away have been detected. Such huge numbers can seem almost meaningless.

Where are we?

The Earth is one of eight major **planets** which orbit about a **star** called the Sun. The Earth has one natural **satellite** called the Moon; the other planets have various numbers of satellites, from none to more than 17. Many of these are no more than small rocks. Some satellites have moons of their own. There are also several thousand minor planets including the 'dwarf planet' Pluto, as well as smaller objects called **asteroids**, orbiting the Sun.

Resource: **http://solarsystem.nasa.gov/planets/profile.cfm?Object=Dwarf**

All of these objects combined make up the **Solar System** (the Sun is also known as Sol). The Solar System is one of millions of star systems that make up a **galaxy** called the Milky Way. Thousands of such galaxies have been detected from the Earth. The sheer numbers of each type of astronomical body are awe-inspiring; the sizes, distances and time periods involved are almost beyond our comprehension.

There are certain numerical facts that we know about the Earth, but we don't usually appreciate their significance. The Earth is nearly a sphere, with a diameter of 12,756km. Mount Everest is 8850m tall. The Mariana Trench (in the Pacific floor) is 10,911m deep. Compare the size of the Earth with the high and low points on its surface. The Earth's crust or surface, on which most human activity takes place, varies in height by 20km – that's 0.15% of the diameter. Relatively speaking, the skin on a bowl of custard is thicker – and a snooker ball is less smooth! You will find the 'relief features' shown on some globes to be greatly exaggerated.

The Sun

The Sun is a star, much like many of those you can see at night. The only reason why it looks so much bigger is that it is so much closer to Earth than the other stars (many of the visible stars are, in fact, considerably bigger than the Sun). The Sun is a giant ball of gases (mainly hydrogen), almost 1.4 million kilometres in diameter (nearly 110 times the diameter of the Earth). It contains over 99% of the matter in the Solar System.

Because of its great size and mass, the Sun's gravity is extremely intense. The gravity squashes the molecules of hydrogen in the Sun, so that they are so tightly packed together and at such high pressure that the temperature approaches 15,000,000°C at the core – enough to make the atoms of hydrogen fuse together to form helium. This process, known as nuclear fusion, releases vast amounts of energy as heat, light and other forms of radiation. This reaction has been going on for the last 5000 million years, and is likely to continue for the next 5000 million. When the Sun finally runs out of hydrogen, it will cool down and expand into a red giant as the internal nuclear reactions and the gravitational attraction reach a new balance. This new version of the Sun will be so large that the Earth's orbit would be inside its circumference – but the Earth will be long gone by this time.

The Sun is the only source of light in the Solar System. Objects such as the Moon reflect the light of the Sun, allowing us to see them. The stars that we see at night are well beyond the limits of the Solar System. It is interesting to note that other star systems have more than one 'sun', or star. Our nearest 'neighbouring' star system, Alpha Centauri, contains three stars which orbit each other. It is the gravitational pull of the Sun, combined with the gravitational attraction of the individual planets, that keeps the planets in their orbits. (See Chapter 4, page 85.)

NB Remind the children never to look directly at the Sun. Doing so can cause permanent damage to the retina of the eye (see page 106). They can use shadows, where necessary, to determine the Sun's position.

The planets

There are eight planets in orbit around the Sun. In order of distance from the Sun, they are: Mercury, Venus, Earth, Mars,

Jupiter, Saturn, Uranus and Neptune. The first four, 'the inner planets', can be described as 'rocky'. The next four, 'the outer planets', can be described as 'gaseous'. Pluto, once the ninth planet, is a bit of an anomaly: a small, rocky planet, less than a quarter of the diameter of Earth, is one of an increasing number of 'dwarf planets' orbiting beyond Neptune.

The further out from the Sun a planet is, the longer it takes to complete an orbit. The duration of the inner planets' orbits can be measured in Earth days: 88 days for Mercury, 687 for Mars. (For a complete listing of such data, please refer to the ASE publication *Primary Signs and Symbols*.) The orbits of the outer planets can best be measured in Earth years: from 11.9 years (11 years and 10 months) for Jupiter, to 165 years for Neptune.

Another difference between the planets is temperature: those planets further out from the Sun are colder. However, the thickness of a planet's atmosphere and the time that it takes to rotate on its **axis** both have an appreciable effect on the inner planets. Mercury, with a fairly minimal atmosphere and long day or period of axial rotation (nearly 60 Earth days), has a significant temperature difference between the daylight and night sides. Each 'day' lasts up to 88 Earth days. Venus, on the other hand, has a very thick atmosphere that collects and retains heat from the Sun very effectively: the surface has an average temperature of 500°C. It is also peculiar in that the Sun rises in the west and sets in the east (115 Earth days later). It is likely that the close proximity of the Sun is the main cause of the relatively slow rotation of these two inner planets.

Both Earth and Mars have atmospheres. They have roughly the same day length; but Mars is, on average, nearly half as far again from the Sun as Earth, and takes almost twice as long to orbit.

Of the 'gas giants', Jupiter is the biggest with a diameter of nearly 144,000km. Jupiter, Saturn, Uranus and Neptune all have 'rings': bands of orbiting debris.

The Moon

Other planets have satellites, and some have many; but none is as large relative to the planet as the Moon is to the Earth. The Moon can only been seen where it is illuminated by the Sun; as the relative positions of the Earth, Moon and Sun change, the amount of the Moon that can be seen from the Earth changes (see below). The Moon orbits the Earth at an approximate distance of 384,000km and is much closer to Earth than any of the planets or the Sun. Its diameter is nearly a quarter that of the Earth (3476km), but is much less dense and so has a much smaller gravitational attraction than that of Earth. It has no atmosphere and no life. So far, it is the only body in space that humans have visited (the first Moon landing was in 1969).

Of the various planetary objects that we can see in the night sky, the Moon appears the biggest; but it is by far the smallest (it's just much closer). Its size and position are such that, when it moves directly between the Earth and the Sun, it can completely block the light from the Sun for a short period and cast a shadow: a solar eclipse, as happened in parts of the UK in 1999. A solar eclipse can be likened to you looking out of the window and blocking your view of a distant house by placing your thumb in the way. The Moon is in such a close orbit (taking approximately 27.3 days) that the same side (or face) of the Moon is always directed towards us.

Asteroids

When our Solar System was forming, not all of the material orbiting the Sun came together to form large planets. As well as the eight major planets, there are thousands of minor planets in orbit around the Sun. Dwarf planets vary in size from over 2000km in diameter (Pluto and Eris) to 950km (Ceres). Asteroids are smaller, right down to the size of grains of sand. It is possible that some may even be debris from the break-up of a larger planet in pre-human times, but this is not certain. Although most known asteroids in our Solar System orbit the Sun between the orbits of Mars and Jupiter (the 'asteroid belt'), a significant number orbit further out and more orbit closer in. Some of the asteroids have very eccentric orbits: Icarus, for example, varies from being closer to the Sun than Mercury to being further out than Mars. Several asteroids cross the path of Earth. The largest of the dwarf planets, Eris, can be up to 97 times as far away from the Sun as the Earth.

Comets

According to Patrick Moore, comets are 'dirty snowballs', mostly being made of loosely packed ice, frozen gases and rocks. Generally, comets have the most eccentric orbits of all. The tail of a comet is not an effect of rushing through space: it is caused by gas particles in the comet being heated as it approaches the Sun. These gases are then blasted off into space by the solar wind: a stream of charged particles (electrons and protons) produced as a result of the nuclear fusion within the Sun.

Halley's Comet is possibly the most famous comet. It orbits the Sun every 75 to 79 years. It was identified by Chinese astronomers in 239BC, but was named after the British astronomer Edmond Halley when he predicted its return in 1758. It passed close to the Sun twice in the 20th century (1909 and 1986), and is next due in 2061. In 1986, an unmanned spacecraft visited it and discovered that it had a solid core approximately 10km in diameter.

Stars and constellations

The Sun is much smaller than many other stars that can be seen from Earth – such as Betelgeuse, a red giant star that is found at the 'right shoulder' of the constellation of Orion. If you replaced the Sun with Betelgeuse, Mercury would be inside the star, and so would Venus and Earth. Mars would be skimming around the surface (if it maintained its current orbit). Other stars are much smaller: one white dwarf appears to be only 3500km in diameter. The size of a star depends on how much matter (mass) it started out with and its age. The size determines the heat (energy output) of the star, which affects its colour. The larger the star, the cooler (and hence redder) it tends to be. Small stars tend to be white hot.

Stars have a life cycle that lasts through several phases. Stars are 'born' in gaseous nebulae when enough gas particles come together to begin to form a gravitational field and draw in more particles. When enough mass has been gathered, nuclear fusion begins. For a star the size of our Sun, this phase will probably last about ten billion years – we are halfway through! A star which has five to ten times the mass of our Sun will go through this phase much faster, in 15 to 50 million years. The end of the nuclear fusion phase is reached when all of the hydrogen in the star has been turned into helium.

Now further reactions take place: the helium is fused to produce carbon, and the carbon is fused to produce still heavier elements. By this time, the cooling star has expanded into a red giant. A star with the mass of the Sun will then shed its outer layers and collapse to become a white dwarf. A much larger star will explode into a supernova and collapse to form a very dense body: either a neutron star (all the mass of the Sun would fit into a neutron star 30km in diameter) or a black hole (the gravity of which is so strong that even light cannot escape from it).

A star is often described as belonging to a particular **constellation**: an imaginary shape made by linking stars (as in a dot-to-dot picture) in a particular part of the sky. Often these stars are nowhere near each other in terms of distance from the Earth, but they are in the same direction when viewed by us. Many of these star patterns in the sky of the northern hemisphere were named by the Sumerians (c4000BC), whereas those seen from the southern hemisphere were named by European explorers in the 16th and 17th centuries. The original constellation names probably had a religious significance. You may well have difficulty in making out the constellations, because of lights nearby and smoke or dust in the atmosphere.

Why you need to know these facts

The subject of astronomy continually fascinates children. This is an area that really allows their minds and imaginations to soar, and it may sometimes be necessary to inject a little reality – though the truth is often more amazing to them than anything they have imagined!

Vocabulary

Asteroid – an irregular rock in orbit, too small to be called a planet.
Axis – in imaginary line going through the centre of a body. Most bodies in space tend to rotate around an axis.
Constellation – an artificial grouping of stars that appear in the same part of the sky.
Day – the length of time a body takes to rotate on its axis.
Dwarf planet – a minor planet.
Galaxy – a large collection of star systems (such as the Milky Way).
Orbit – the regular path that one body in space takes around another when under its gravitational influence.
Planet – a non-luminous body that orbits a star (such as the Earth).
Satellite – a body that orbits a planet (such as the Moon).

Solar System – the name given to the Sun and all the bodies in orbit around it.
Star – a luminous body in space (such as the Sun).
Year – the time taken to complete an orbit. *(See page 160)*.

Amazing facts

- Because the Moon is in a close orbit around the Earth, the same side of the Moon is always directed towards us. We cannot see the 'dark side'.
- Until the Russians sent an unmanned space craft around the Moon in 1959, some people believed that 'moon people' were living there.
- Although the Moon appears bright to us, only 7% of the sunlight falling on its surface is reflected back into space – meaning that it reflects about the same proportion of light as coal dust.
- If we imagine the Earth to be a marble (1.5cm in diameter), the Sun would be a ball about 1.75m in diameter. To be at the correct relative distance, they would need to be placed 190m apart. Feel the heat of the Sun on a summer's day, and think about the total energy being generated!
- The planet Saturn is less dense than water – if you could find a big enough bowl of water to put it in, you could make Saturn float.
- One of the brightest 'stars' in the sky isn't a star at all, but the planet Venus. Although it does not give off any light of its own, it appears so bright because its covering of white clouds reflects light from the Sun very effectively.
- Concorde was the fastest passenger aeroplane we have yet invented. If it could fly in space, a Concorde jet aircraft would take nearly seven days to get to the Moon – and over seven years to reach the Sun.

Common misconceptions

The Earth is flat.

This idea is related to our experience of gravity: if the Earth is a sphere, why don't people on the bottom drop off? The key here is to gradually change the child's view and use of terminology so that he or she begins to appreciate that gravity acts towards the centre of the Earth – that is, that the meaning of 'down' is relative to our position on the Earth.

Questions

Where do the stars go to during the day?

They are still there, but the Sun is too bright for us to be able to see them. I usually demonstrate this by setting up a few lit 2.5V and 3.5V bulbs (powered by 1.5V batteries) in non-obvious places around the room on a bright, sunny day. I bring the children in, sit them down and ask them whether they notice anything unusual. I then draw the curtains (the more opaque the better). At this point, the brighter bulbs might be noticed. I then turn off the lights, so that the small bulbs are the only light source in the room.

Why do stars twinkle?

Stars do not flicker on and off in reality. They appear to because of atmospheric pollution. Particles of dust or smoke floating in the atmosphere block some of the light coming from a star and so make it appear to twinkle. If you could view the stars from beyond our atmosphere, there would be no twinkling effect – and you would be able to see a lot more of them! The light reflected from planets will also appear to twinkle; but the Moon's image, which is much larger, does not twinkle because the particles are not large enough to produce momentary variations in its brightness as they drift past.

Is there life on any other planet?

Quite probably, but not in this Solar System. There is probably life on other planets orbiting other stars, but it may well be very different from what we know as life here. Over the past few years a number of planets have been discovered orbiting other stars, some even 'Earth-like' in their size and distance from their star.

Teaching ideas

Star gazers (modelling, making)

Children can copy constellations from a star chart (or design their own) by making pin-prick holes in tinfoil, then stretch the tinfoil over the end of a cardboard tube to make a 'telescope'. Alternatively, the 'star' holes can be cut in large sheets of black sugar paper (bigger holes for brighter stars), and the sheets stuck to the classroom windows.

Virtual astronomy

There are increasing numbers of websites that allow you (and children) to view star maps of the night sky and follow the motion of planets.
Resources:
www.bbc.co.uk/science/space/
www.nasa.gov/multimedia/3d_resources/assets/tycho8.html
http://neave.com/planetarium/

Concept 2: Days, months and years

Subject facts

Day length

A **day** is the time it takes the Earth to spin on its axis. This is quite a difficult idea for us to comprehend: the Sun stays still and the Earth rotates. It was not until the 16th century that evidence for this view was successfully gathered.

The apparent movement of the Sun across the sky is regular and predictable. On the equator, the Sun is directly overhead at noon (Greenwich Mean Time, not British Summer Time) on the days of the spring and autumn **equinoxes**. In the northern hemisphere, the Sun will appear slightly to the north at noon in summer and slightly to the south at noon in winter. The Sun's apparent height varies within the day (it is at its peak, or zenith, at midday), and on a seasonal basis (see below). If you plot the motion of the Sun using a shadow stick, the shadow will be shortest at midday and will move 15° in each hour (360° in 24 hours).

Phases of the Moon

The **calendar month** is a human attempt to make the number of months (12) fit exactly into 365 days. A **lunar month** is the time it takes the Moon to complete a cycle of phases: 29.3 days. The Moon orbits the Earth in 27 days 8 hours; the additional time is due to the movement of the Earth about the Sun (see Figure 1).

Figure 1 Paths of the Earth and Moon

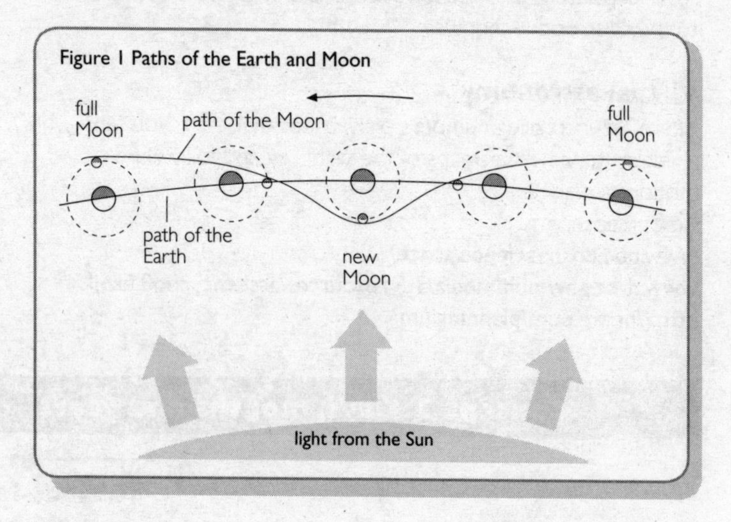

The phases of the Moon (see Figure 2) occur because we can only see the portion of the Moon that is illuminated by the Sun. When the Sun, Earth and Moon are in line (in that order), won't the Moon be in the Earth's shadow? Sometimes this does happen (a lunar eclipse); but the Moon is normally slightly above or below the plane of the Earth's orbit.
Resource: **www.history.co.uk/videos.html?bctid=23646683001& Phases-of-the-Moon-**

During a full Moon as seen from the Earth's equator, the Moon will appear to rise just as the Sun sets, be at its zenith (highest point) at midnight and set as the Sun rises. If the Moon moves on a further seven days in its orbit (the Sun, Earth and Moon should form a 90° angle), half of the Moon will be visible from the Earth and the other half will be in darkness (last quarter). At dawn, the Moon will be high in the sky and will be

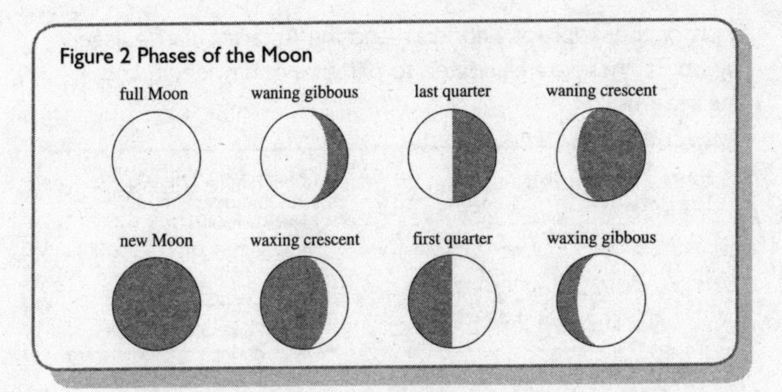

Figure 2 Phases of the Moon

visible until the Sun becomes too bright – at midday, when the Moon sets, it probably won't be seen anyway.

Over the next week, the visible portion of the Moon will become smaller and smaller (crescent Moon) and will appear in the sky closer and closer to the Sun (just before dawn). By the time the Sun, Moon and Earth line up (in that order), the Moon will not be seen in the sky because all of its illuminated side will be pointing away from the Earth. Sometimes, in this phase, the Sun, Moon and Earth form an exact line and the Moon casts a shadow on the Earth: a solar eclipse.

The first visible sliver of the Moon (new Moon) will appear just after dusk a day or so later, very close to the horizon. The illuminated portion of the Moon, as viewed from the Earth, will get larger and larger until the whole right-hand side of the Moon is visible (first quarter). At this point, the Sun, Earth and Moon once more form a 90° angle. As the Moon moves towards being full once more, it will go through a gibbous phase where a crescent appears to have been 'cut out' of it.

Understanding the globe

Have you ever wondered why globe maps are mounted on stands at an angle? It is not so that the British Isles are 'easy to view', but so that the Earth's axis remains at a particular angle (23.5°) to the plane of its orbit. This axial inclination of the Earth means that the North Pole always points towards the same region of space. Currently this is the star Polaris, which can be found in the tail of the constellation Ursa Minor (the 'Little Bear').

There are five lines drawn around the globe map. From North to South, these are: the Arctic Circle, the Tropic of Cancer, the equator, the Tropic of Capricorn and the Antarctic Circle, (see Figure 3). These are all related to patterns of day length and the seasons.

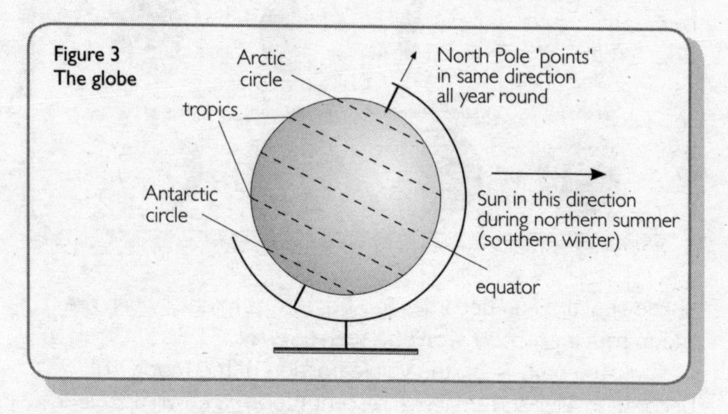

Figure 3
The globe

Equinoxes and solstices

Seen from the equator, the Sun will be directly overhead at midday on two days of the year: the spring and autumn equinoxes, usually 20 March and 22 September (points C and D in Figure 4). On these two days, wherever you are on the Earth (apart from inside the Arctic and Antarctic Circles), you will receive 12 hours of daylight and 12 hours of night. On these days only, the Sun will rise directly due east and set directly due west.

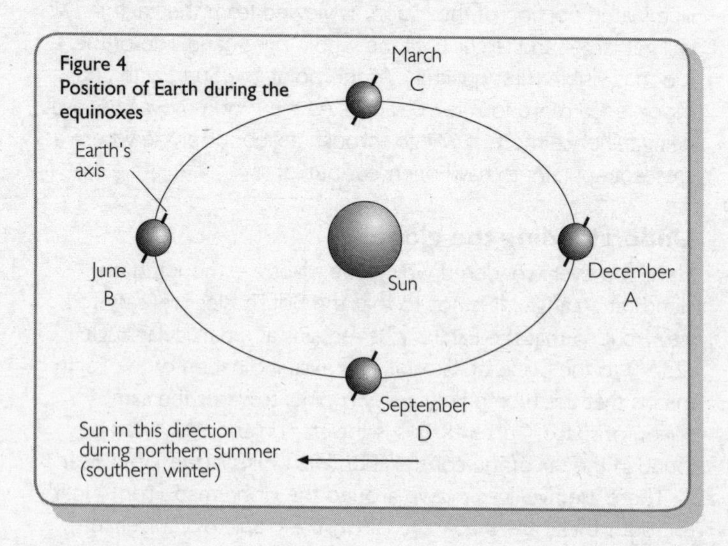

Figure 4
Position of Earth during the equinoxes

In the northern hemisphere, the longest 'day' (duration of daylight) is 20 June and the shortest 22 December. These are known as the summer and winter **solstices**. The reverse is the case in the southern hemisphere. Throughout the summer in Britain (between the spring and autumn equinoxes), the Sun will rise in the north-east and set in the north-west. It will rise and set in the most northerly points at the summer solstice. In the winter, it will rise in the south-east and set in the south-west, reaching the most southerly points at the winter solstice.

These changes affect the times of sunrise and sunset. Figure 5 shows the position of the Sun at different times of the day (all GMT) for different times of the year. In Britain's summer months, the days are longer than the nights; in the winter months, the nights are longer than the days.

Figure 5 Position of the Sun throughout the year

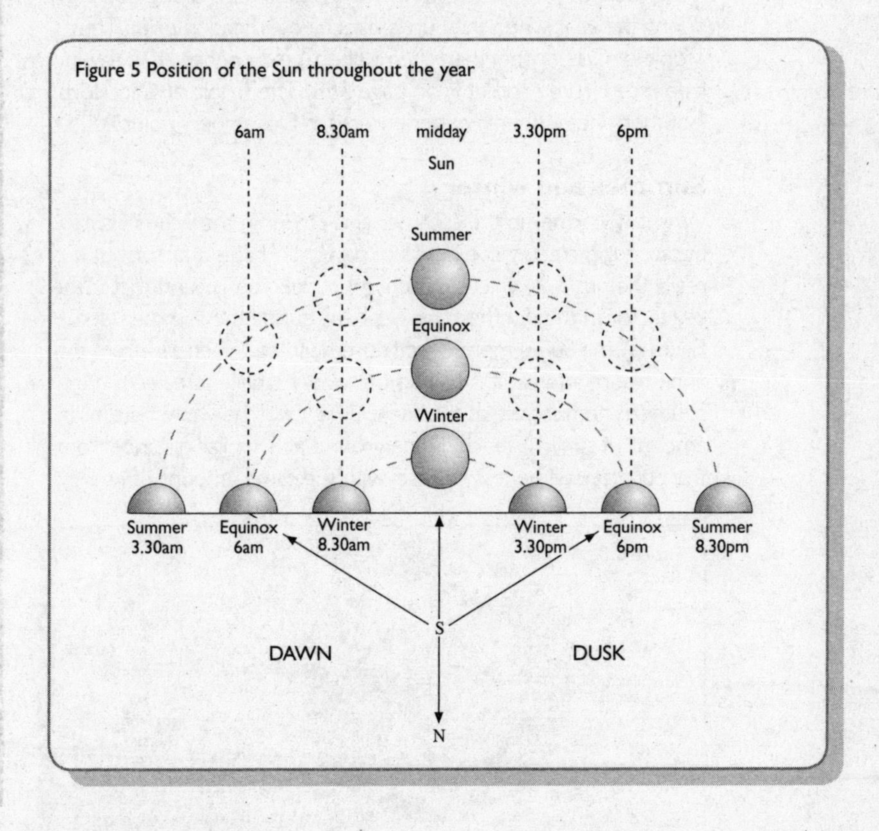

Polar days and nights

The polar regions (within the Arctic and Antarctic Circles) are an exception to this pattern. In the polar regions, the Sun will make its first appearance of the year at the spring equinox. It will appear to do a complete circuit on the horizon. During the 'summer', the Sun will continue to circle without setting for six months. It will reach a 'high point' in the middle of the summer, and very slowly spiral back down. Once it has set, it stays set for six months. So in effect, the poles have two 'days' per year. There are many videos available showing the Arctic winter sun phenomenon on websites such as Youtube®.

Tropical sunshine

From the viewpoint of either of the tropics, there is one day during the year when the Sun is directly overhead at noon. The Tropic of Cancer (northern hemisphere) experiences this day in June (see Figure 4, point B on page 158). The Tropic of Capricorn (southern hemisphere) experiences it in December (point A).

Summer and winter

Why is it warmer in the summer and colder in the winter? Is it because there are more hours of daylight? That is a factor, but it is really the *angle* at which the sunlight strikes the ground that is the key factor. In the northern hemisphere during the summer (see Figure 6), the Sun is more directly overhead, so the energy from the Sun is more intense: it is spread out over a smaller area (A). In the southern hemisphere at the same time, it will be winter: a similar amount of sunlight reaching the ground at a similar distance from the equator will be less intense, with the same amount of energy

Figure 6

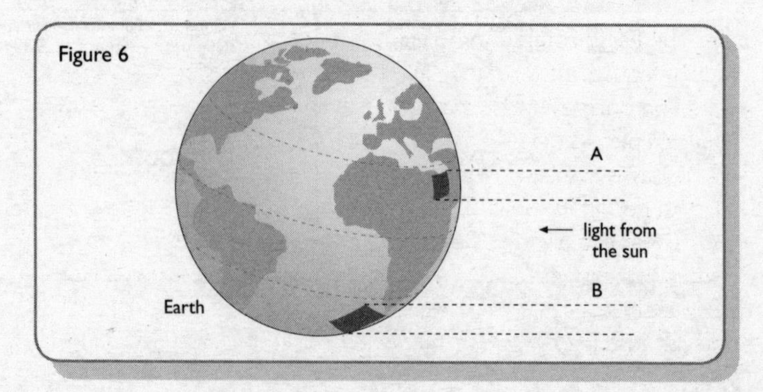

being spread over a wider area (B). These seasonal differences are crucial to the lives of plants, which respond to the changes in day length and temperature by entering new phases of their life cycle.

Why you need to know these facts

The most significant time periods in common use – the day, month and year – are based on astronomical observations. Others, such as the hour and week, are derived from these basic units. Understanding the planetary motions on which our time units are based is a core concept in science.

Vocabulary

Day – the period the Earth, or other body, takes to rotate once on its axis.
Equinoxes – days when the periods of daylight and night are equal (March and September).

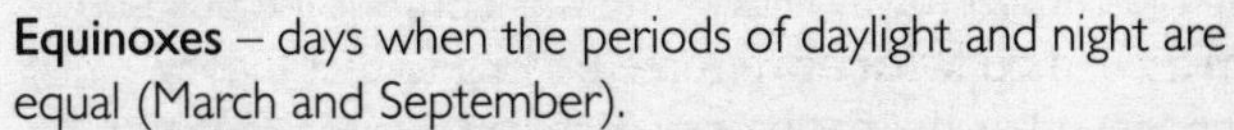

Lunar month – the period the Moon takes to complete one orbit of the Earth (not quite the same as a **calendar month**).
Solstices – the shortest and longest days of the year in terms of the period of daylight (June and December).
Year – the period the Earth takes to complete one orbit of the Sun.

Amazing facts

- The Earth rotates approximately once every 24 hours – so at the equator, it is rotating at 460 metres per second. It orbits the Sun once every year, travelling at an average speed of 30,000 metres per second.
- A day is actually 23 hours and 56 minutes. It varies slightly in length during the year.
- A year actually lasts 365 days, 5 hours, 48 minutes and 46 seconds. Leap years are necessary to allow for the extra quarter-day each year.
- Because leap years were not being used accurately enough

(it is not *exactly* an extra quarter-day each year), in 1752 the calendar was adjusted so that the next day after 2 September was 14 September. This caused riots because people thought they were being robbed of 11 days of their lives!

Common misconceptions

Night and day

Children have various ways of explaining what happens to the Sun at night. Some ideas are very naive, such as 'It goes behind the hills.' This is really indicative of a 'flat Earth' concept, and needs to be challenged with stimuli that represent the Earth as a sphere in space.

The idea that the Sun 'goes behind clouds' can be challenged with observations on a cloudy day: the Sun is not visible, but it's not night. This could be countered with the claim that these are 'not the right clouds'; observations of the passage of the Sun during the day will help to clarify the situation, but the build-up of clouds around dusk does give some backing to that theory.

'The Sun orbits the Earth once each day' was the most commonly held view for thousands of years. It was only close observation of other planets, and the need to explain their movements, that led astronomers (notably Copernicus, Galileo and Kepler) to reject it in the 16th century. Children sometimes believe that the Earth orbits the Sun once each day. Both of these ideas can be challenged by using role play to explore planetary movements, or by demonstrating with a globe and a light source (see Teaching ideas below).

Seasonal changes

All children are aware of the seasons; but because of the timescale relative to their own experience, the pattern may be difficult for them to appreciate. Children living in cities may find seasonal changes difficult to identify at times. Some children may reverse the link with changes in nature, believing that trees losing their leaves causes winter. Other suggestions might be as follows:

- 'We are further away from the Sun in winter.' (We are actually slightly closer.) Challenge them by drawing their attention to Australian soap operas where Christmas comes in the middle of summer.

- 'The Earth goes around to the cold side of the Sun.' This is quite a nice one, but the Australian summer example still helps.
- 'A cold planet comes close and takes the heat away.'

This is surprisingly difficult to argue with, since it is based on pure imagination. The correct idea that it has to do with the angle of impact of the Sun's rays on the surface of the Earth can be modelled using a narrow-beam torch and a globe (see above).

Phases of the Moon

Children's explanations of the Moon's phases range from the very naïve to the plausible but wrong! Examples include:

- Clouds get in the way. Challenge this by asking: *Why are the clouds exactly the right shape? How can you see a crescent Moon on a cloudless night?* A lack of focused observation is probably behind this idea.
- The shadow of the Earth, another planet or the Sun causes the shape. Say: *That might cause us to see a crescent-shaped Moon, but what about a quarter Moon or a gibbous Moon?*

The correct explanation can be demonstrated using a light-coloured ball on a stick and a good light source, such as an angle poise lamp, in a darkened room (see Teaching ideas below). Encourage the children to observe and remember the position of the 'Moon' relative to the 'Sun' at each point, and to use this to explain the phases.

Questions

Why do we have leap years?

Because the Earth doesn't go around the Sun in an exact number of days. All of those bits of days add up. If we ignored them, in a few hundred years we would be having our summer holidays in the middle of winter!

Teaching ideas

For the most part, all you can do to teach this area of science is to demonstrate and model.

Rotation of the Earth (modelling, role play)

This process can be acted out by a group of five children. Start by asking them to imagine standing with their backs to a weather map: their feet are south, head north, left hand east and right hand west.

Now one child takes the (somewhat inactive) role of the Sun, and the other four represent the Earth at different times. The four 'Earths' link arms to form a circle, facing outward (see Figure 7). One of them faces the 'Sun'. The ones on either side have to look over their shoulders to see the 'Sun', and the one round the back can't see the 'Sun' at all. For the child facing the 'Sun' it is midday, for the one who can't see the 'Sun' it is midnight, and for the other two it must be dawn and dusk – but which is which? Remind them that left is east and that the Sun rises in the east – so the child with the 'Sun' to his or her left must be dawn. Logically, therefore, the remaining child must be 'dusk'.

Figure 7 Modelling the Earth and the Sun

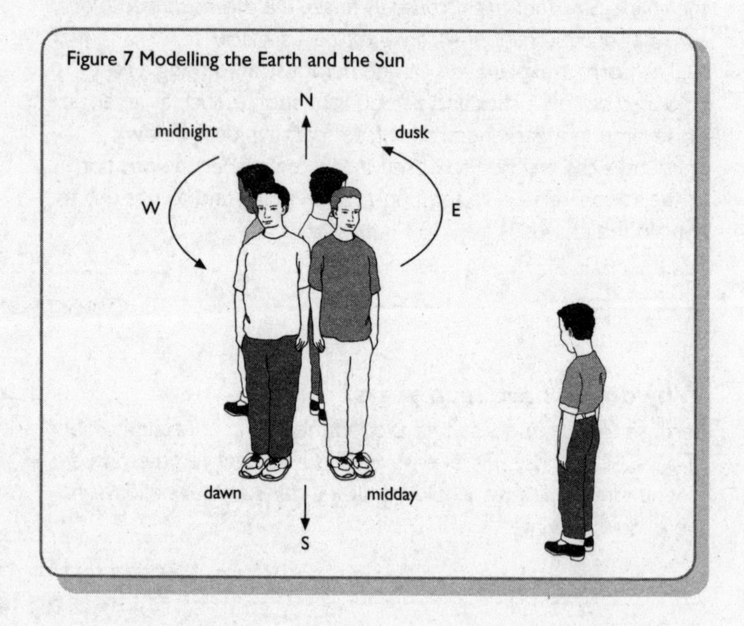

Assume that dawn, midday, dusk and midnight are each six hours apart. If the day were to move on by six hours, dawn would become midday and so on. So which way must the Earth be spinning when you look down from above the North Pole? (Anti-clockwise). The children can use role play to work this out, perhaps starting with six hours (to get the hang of it) before moving on to 24 hours. They can decide whether it is night or day by whether they are able to see the 'Sun'.

Phases of the Moon (modelling, role play)

The children can use a similar role play method to explore the phases of the Moon (see Figure 8). Alternatively, you can demonstrate the cycle as follows. Go into a blacked-out classroom with a lamp and a ball on a stick (a large polystyrene or foam one would be best). Place the lamp at one end of the room with the children in the middle, leaving you room to walk around them. If the lamp or the walls are too bright, there may be too much light in the room to show light and dark sides of the ball. When the ball is between the lamp and the children, they cannot see the lit side of the ball (new Moon). Moving anti-clockwise, first a thin crescent, then a first quarter, then a gibbous ball can be seen before you reach the far side of the lamp and the children to display a full Moon. Continue round to complete the cycle.

Figure 8 Modelling the phases of the Moon

Seasonal change (teacher demonstration)

Use a narrow-beam torch and a globe map to demonstrate that in the winter, due to the inclination of the Earth's axis, the light from the Sun will be spread over a greater area. Because the Sun's energy is being spread more thinly, the temperature will be lower.

Shadows (observing, measuring, using ICT)

The children can use shadows to observe the apparent position and movement of the Sun. They should use the same pole or stick each time, placed in the same position on flat ground, to obtain a shadow. They can record the length and direction of the shadow at different times in the day. The changes from day to day will be minimal; but if the children record every day on a weekly rota, they will soon build up a considerable bank of data. From midwinter to midsummer, the shadows will become progressively shorter for any given time of the day (watch out for the switch to BST!), but will remain in the same direction. The data can be recorded on a spreadsheet and presented in the form of stick charts (bar line graphs).

Resources

For much of the classroom-based work on this area of science, the main requirement is for a strong light source (data projector, focused theatre lamp or even a torch) and a blackout room.

Beyond the basic practical explanations, further information about the Solar System and the Universe can be obtained through research.

Websites

www.bbc.co.uk/science/space/
www.nasa.gov/topics/solarsystem/index.html
www.bis.gov.uk/ukspaceagency/discover-and-learn
www.esa.int/esaSC/SEM4PZVJD1E_index_0_ov.html
http://youtu.be/77ZoF7Y1pNk

Books and CD-ROMs

Scholastic Primary Science: Earth Watch
Scholastic Primary Science: Out of this World
Investigate: In the Sky (Scholastic) – exciting non-fiction readers

Glossary

Electricity

Battery – a case containing chemicals that react to cause a flow of electrons when connected into a complete circuit.
Bulb (lamp) – a device which, when connected into a circuit, will resist the flow of electricity and heat up, producing light.
Circuit – a complete path between the two terminals of a battery, made with materials that conduct electricity.
Conductor – a material that allows the free flow of electricity through it (that is, the free movement of electrons between atoms).
Current – the rate of flow of electricity, measured in amps (A).
Motor (electrical) – a device for changing electricity into movement through electromagnetic induction.
Parallel circuit – where there is more than one path that the flow of electricity can take to complete a circuit.
Power – how much energy is used in a given time, measured in joules per second or watts (W).
Resistor – a device that resists the flow of electricity in a completed circuit. Some resistors, such as a bulb or a motor, use the 'friction' of the restricted flow to produce light, heat or movement.
Series circuit – where there is only one path for the flow of electricity to take, through all the components in the circuit.
Switch – a device for controlling the flow of electricity in a circuit.
Voltage – the difference in potential between two parts of a circuit, 'pushing' the current. Voltage is measured in volts (V).

Magnetism

Ferromagnetic material – a type of metal (iron, steel, cobalt or

nickel) that can be strongly magnetised.
Hard magnetic material – can be permanently magnetised.
Magnet – an object that produces a magnetic field.
Magnetic compass – a free-floating magnet which aligns itself with the Earth's magnetic field.
Magnetic material – a material that is affected by a magnetic field.
Magnetic north and magnetic south – the points on the Earth's surface where the lines of force in the magnetic field are vertical.
North-seeking pole – the pole of a magnet that aligns itself to the Earth's magnetic north pole.
Soft magnetic material – can be temporarily magnetised.
South-seeking pole – the opposite of a north-seeking pole.

Energy

Conservation – the process of reducing our overall energy use and needs by making sure that energy is used more efficiently.
Energy – the capacity to do work.
Energy transformer – something that transforms one form of energy into another, as our bodies turn chemical energy from food into movement, heat and sound.
Engine – a device for turning stored energy into useful work or movement.

Forces

Accelerate/decelerate – to increase or reduce speed of movement in a particular direction.
Balanced forces – where the forces acting on an object are equal and opposite, causing no change in the object's speed, direction or shape.
Density – the mass of an object per unit of its volume.
Drag – the frictional force experienced by an object moving through a fluid.
Force – a push or pull.
Friction – a force between two surfaces that acts against the direction of movement.

Gear – two wheels with serrated or notched rims that mesh together to transfer movement.
Gravity – the force of attraction between two masses. The Earth attracts objects on its surface with a force of 9.8N per kilogram of the object's mass.
Lever – usually a rigid bar with a pivot point close to one end, allowing a large movement at one end of the lever to be converted into a smaller movement at the other, which effectively magnifies the force applied.
Mass – the amount of 'stuff' or material in an object (measured in kilograms or kg).
Newton (N) – the unit of measure for force.
Pressure – force applied per unit area.
Pulley – a wheel with a grooved rim that allows the transfer of movement via a belt or band.
Streamlined – a shape which minimises the profile presented by an object in order to minimise the resistance it encounters when moving through a liquid or gas.
Unbalanced forces – where the forces acting on an object are greater in one direction that in another, causing a change in its speed, direction or shape.
Upthrust or **buoyancy** – the upward force exerted on a body by a fluid that surrounds it, equal and opposite to the weight of the water displaced.
Weight – the force exerted on the ground by a mass due to gravity.

Light

Absorption – when light strikes a surface and is retained within it.
After-image – negative image 'seen' as a result of parts of the retina closing down after overexposure to light or particular colours of light.
Cornea – the tough outer covering of the eye.
Darkness – the absence of light.
Iris – opens and closes to let more or less light into the eye.
Luminous – able to produce light.
Opaque – a material which blocks the passage of light.

Optic lens – the clear opening that allows light into the eye.
Optic nerve – transmits the information received by the retina to the brain.
Primary colours (red, blue and green) – the colours of light that our eyes are able to detect.
Reflection – when an image is returned from the surface of an object.
Refraction – the 'bending' of light when it passes from one transparent material to another.
Retina – the internal surface of the eye that senses light.
Scattering – when light is returned from a surface.
Translucent – a material through which you can see light, but not an image.
Transparent – a material through which you can see an image.
White light – a mixture of all the colours of light (for our eyesight, we only need red, blue and green to make white).

Sound

Amplitude – the strength or intensity of sound vibrations, commonly perceived as 'loudness' or 'volume'.
Cochlea – the sound reception part of the inner ear.
Eardrum – the membrane which receives sound from the pinna and passes it to the inner ear.
Echo – a sound reflection from a rigid surface.
Frequency – the number of vibrations per second corresponding to a given sound (its 'pitch'), measured in hertz (H2).
Oscilloscope – a device for displaying in a visible form sound waves that have been converted into electrical pulses.
Pinna – the outer portion of the ear.
Transmission – the process by which sound travels from one place to another.

Timbre – the collection of secondary notes that accompany the main note to add richness to the sound.
Vibration – forward and backward movement of an object (usually very rapid).

Earth in space

Asteroid – an irregular rock in orbit, too small to be called a planet.
Axis – in imaginary line going through the centre of a body. Most bodies in space tend to rotate around an axis.
Constellation – an artificial grouping of stars that appear in the same part of the sky.
Day – the length of time a body takes to rotate on its axis, or the period the Earth takes to rotate once on its axis.
Dwarf planet – a minor planet, roughly spherical in shape, which has not cleared its neighbouring region of objects.
Equinoxes – days when the periods of daylight and night are equal (March and September).
Galaxy – a large collection of star systems (such as the Milky Way).
Lunar month – the period the Moon takes to complete one orbit of the Earth (not exactly equal to a **calendar month**).
Orbit – the regular path that one body in space takes around another when under its gravitational influence.
Planet – a non-luminous body that orbits a star (such as the Earth).
Satellite – a body that orbits a planet (such as the Moon).
Solar System – the name given to the Sun and all the bodies in orbit around it.
Solstices – the shortest and longest days of the year in terms of the period of daylight (June and December).
Star – a luminous body in space (such as the Sun).
Year – the time taken to complete an orbit, or the period the Earth takes to complete one orbit of the Sun.

Index

absorption 103, 109–10, 116, 122
acceleration 67, 68, 70, 75–6
after-images 127, 128
air resistance 73, 74, 83, 86–7
amplitude 132, 137, 138
Archimedes 92, 98
axis, Earth 145, 152, 157–8

batteries 10–11, 14, 15, 21–3, 25–6, 31, 39
  rechargeable 20
bulb holders 22–3, 31, 39
bulbs 10–11, 21–3, 25–6, 27–8, 31, 36–7, 39
buoyancy (see upthrust)
buzzers 11, 25, 29–30, 31, 39

calories 58–9
circuits 21–38
colours 103, 105, 120–5, 129–30
compasses, magnetic 42, 52–3, 54
conductors
  electrical 11, 12, 13, 15
  sound 133–4
constructing 32–3, 37, 54, 79–80, 155
current 11, 18, 19

days 145–6, 152, 155–6, 157–8, 161, 162
deceleration 69, 70, 75–6
demonstrating 107, 118–19, 123–4, 128, 166
density 65, 87, 91–3, 94
design and technology 31, 32, 33, 54
developing vocabulary 72
diffusion (see scattering)
diodes 25, 30
displacement buckets 95–6, 101
dynamos 13, 29

ears 131, 132, 133, 140–3
Earth 145–6, 147, 153–4, 157–8, 161
  magnetic field 41, 44, 51–3
  (see also gravity; tides)
echoes 133, 134, 142–3
eclipses 146, 150, 157
elastic materials 60, 61, 65, 69
electric motors 28–9, 31, 39
electrical circuits 21–38
electrical equipment 10–11, 13, 17, 21, 61
electricity
  generation 13–14, 29
  mains 10–11, 14, 20–1

nature 12–13, 14, 16
safety 10, 17, 20, 21, 40
(see also electrical equipment; batteries; bulbs; circuits)
electromagnets 33, 42, 44, 45, 47
energy 56-63, 69, 110, 112
explaining 26, 72, 85, 108, 120
exploring
forces 71–2, 95, 97
light 108, 120, 123–4
magnetism 48, 51, 54
sound 136, 143
eyes 102, 110, 126–30

ferromagnetic materials 42, 47, 48
floating 65, 88, 92–3, 94
force meters 76–7, 79–80
forces 66–79
balanced 64, 65, 72–5
measuring 76–9
unbalanced 64, 65, 70, 75
(see also floating; friction; gravity; machines; pressure; upthrust)
frequency 132, 138, 139
friction 61–2, 64, 65, 66, 71, 80–5, 87
'static' charges 12–13
fuses 18

gears 99–100
gels (colour filters) 125
globe (Earth) 157–8
gravity 64, 65, 73, 74–5, 76, 85–90, 148

grip 80–1

hearing 131, 132, 133, 135–6, 140–3
heat energy 58, 59, 61

ICT 7–8, 125, 143, 155, 166
identifying, forces 72, 76
investigating
electricity 33
forces 78, 85, 90, 94–5
light 119
magnetism 48–9, 51
sound 139
iron filings 45–6, 51, 55

joules 18, 19, 57, 58

kinetic energy 59, 61

LEDs (light emitting diodes) 25, 30, 39
lenses 103, 108, 115–16, 120, 130
levers 97–8, 100
light 59, 61, 102–3, 104–8, 120–5
absorption 103, 109–10, 116, 122
scattering 102–3, 109–10, 111, 117, 122
(see also colours; eyes; reflection; refraction)
light sources 102, 103–4, 108, 120
lightning 12–13, 15
lodestone 42, 43, 52
loudness 132, 137, 138

lubrication 65, 80

machines 61, 64, 65, 80–1, 97–100
magnetic fields 12, 13, 28, 29, 41, 42, 44, 51
  Earth 41, 44, 52–3
  electric motors 28–29
magnetic forces 44–5, 48
magnetic materials 41, 42, 47, 48–9, 50, 55
magnetic poles, Earth 51–3
magnetism 43, 45, 50–1, 52, 65
magnets 42–8, 50–1, 54, 55
  attraction and repulsion 48–9
  (see also compasses; electromagnets)
mass 65, 68, 71, 83, 87–8, 91
matching 19, 21–2
measuring 79–80, 85, 95–6, 166
mechanisms 97–100
metals 11, 13, 41, 48, 50
mirrors 103, 104, 106, 108–9, 111–12, 120, 130
modelling 23, 26, 132–3, 143, 155, 164–5
months 156, 161
Moon 87, 107, 145–6, 147, 149–50, 153
  phases 146, 156–7, 163
  tides 65, 89–90, 146
motors, electric 28–9, 31, 39
musical instruments 131, 132, 137, 138, 139, 144
musical notes 138

newton meters 76, 101
newtons 65, 66, 76
Newton's Laws of Motion 66, 67, 69, 71, 73–4

observing 51, 53–4
  Earth in space 166
  electricity 26, 31
  forces 71–2, 95, 100
  light 120, 125
  magnetism 50
opaque materials 102, 108–9, 116
orbits 88–9, 152
oscilloscopes 134, 138

parachutes 74–5, 90
parallel circuits 33–6
Petri dishes 46, 51, 55
pigments 123
pitch 132, 137, 139
planets 65, 145–6, 148–9, 152
plastic materials 69
playground equipment 72, 101
polarity, magnets 49
poles
  Earth's magnetic 41, 44, 51, 52–3
  magnets 41, 44, 45, 47, 49
potential energy 59, 61, 70
power, electrical 11–12, 18, 19
predicting 38, 95
pressure 64, 65, 83, 96–7
primary colours of light 103, 121, 123–4
primary science teaching 5–6, 8–9

prisms 103, 120, 123–4, 130
problem solving 32, 33, 38
pulleys 98, 100

rechargeable batteries 20
recording 51, 72, 85, 90, 94–5
reflection
  light 102–3, 104, 111–13, 116, 117, 118, 120
  sound 132
refraction 103, 112–16, 118–19
resistance 11–12
  not useless 13
resistors 13, 15, 33
resonant frequency 139
resources, teaching 38–40, 54–5, 63, 101, 130, 144, 166
role play 164–5
rose-coloured spectacles 122, 125

safety
  Earth in space 148
  electricity 10, 17, 20, 21, 39, 40
  light 106, 108, 116, 129
  magnetism 46, 52, 55
scattering, light 102–3, 107, 109–10, 117, 122
seasons 146, 158–9, 160, 162–3
series circuits 33–6
shadows 102, 105, 109, 117, 155–6, 166
sight 103, 110, 126–30
sinking 65, 91, 93, 94
skydivers 72–3, 74–5
Solar System 146, 147, 148–9, 153
sorting
  electricity 17, 26
  forces 95
  light 108, 119, 125
  magnetism 48
  sound 140
sound 131–2, 137–8
  reception 140–3
  transmission 132–6
space 146–52
spectrum 103, 118–19, 121, 123–4
spring balances 76–7, 101
stars 146, 151–2, 153
'static' charges 12–13
static friction 80
stored energy 59, 62, 70
streamlining 74, 83
Sun 145–6, 147, 148, 150, 153, 156
  energy 56, 59, 62–3
  light source 102, 103
surface area 74, 83
surface tension 81
switches 11, 27, 32, 35–6, 40

teaching, primary science 5–6, 8–9
terminal velocity 74–5
testing
  electricity 17, 26, 31, 38
  forces 84–5, 95–6, 100
  light 125
  magnetism 50

theatre lamps 123, 125, 130
thermal energy 59, 62
thunder 12–13
tides 65, 89–90, 146
timbre 138
tone 138
torches 108, 111, 119, 128, 130
toys 40, 71–72, 101, 144
transformers
  electricity 14
  energy 61, 62
translucent materials 102, 108–9, 117
transparent materials 102, 108–9, 113, 115
tuning forks 136, 138, 144

upthrust 65, 88, 92–3

vibration, sound 131–2, 134, 135
vision 102, 110, 126–30
visual overload 127
voltages 11, 14, 15, 18, 19
volume 91–2, 94
volume (loudness) 137

watts 18
weight 65, 87–8, 91
wineglasses, smashing 139
wire strippers 22, 39
wires 22, 24, 39

years 153, 161